101

Human Resources

QUESTIONS
ANSWERED

Katherine Dobbs, RVT, CVPM, PHR
Louise S. Dunn

AAHA®
press

Veterinary Solutions Series
101 Veterinary Human Resources Questions Answered
© 2013 by Katherine Dobbs, RVT, CVPM, PHR; Louise S. Dunn

press

American Animal Hospital Association Press
12575 West Bayaud Avenue
Lakewood, Colorado 80228
800/252-2242 or 303/986-2800
press.aahanet.org

ISBN 978-1-58326-185-9
Library of Congress Cataloging-in-Publication Data

Dobbs, Katherine.
    101 Veterinary human resources questions answered / Katherine Dobbs, Louise S. Dunn.
        pages cm. -- (Veterinary solutions series)
    One hundred one human resources questions answered
    One hundred and one human resources questions answered
    Includes bibliographical references.
    ISBN 978-1-58326-185-9
    1. Veterinary services--Personnel management. 2. Veterinary services--Administration. 3. Office management. I. Dunn, Louise S., 1955- II. American Animal Hospital Association. III. Title. IV. Title: One hundred one human resources questions answered. V. Title: One hundred and one human resources questions answered. VI. Series: Veterinary solutions series.
    [DNLM: 1. Personnel Management--methods. 2. Veterinary Medicine--organization & administration. 3. Hospitals, Animal. SF 756.4]
    SF756.46.D63 2013
    636.08'32068--dc23
Cover and series design by Erin Johnson Design

Printed in the United States of America

11 12 13 / 1 2 3 4 5 6 7 8 9 10

With thanks to Dr. R. Dennis Heald who taught me the management motto, "It's not your fault, but it IS your problem," and Dr. Jim C. Vulgamott who taught me how to "put 'em on the roof."

KATHERINE DOBBS

With thanks to my husband, Joel Gendelman, and all my family and colleagues who supported and encouraged me.

LOUISE S. DUNN

CONTENTS

PREFACE

Most of us enter veterinary medicine because we love animals. Yet, we soon realize that there is more to veterinary practice than meeting the needs and wants of those furry family members. We come to understand that, in addition to there being a family member at the other end of that leash, we have to serve our practice team members and consider their needs for our businesses to succeed. This book aims to help you create a team that performs well, provides great service to animals and people, and maintains the motivation to serve pets and the families who love them.

It's often said that the animals are the easy ones to treat while the team is something altogether different and, at times, much more complicated. For animals to receive excellent medical treatment, and clients an excellent experience, the practice management and leadership must make a combined and purposeful effort to create a professional and satisfying work environment for team members so they are happy to come to work and do their jobs (and tough jobs these are!), and patients and clients can benefit from their knowledge and skills.

A veterinary practice can only be successful because of the hard work and strong vision of its team members. Those team members do their best when they have guidelines to follow for their roles and the overall functions of the practice. That's where human resources (HR) comes in. There are a number of tools and applications on how to manage the team; these can be used to culminate in an HR department or function of the practice. Even a small practice will benefit from having HR practices in place—and in writing.

While the endeavor to keep the team members happy and excited about their jobs is an HR function, there are also legal considerations for many HR tasks, such as hiring people who may work in the United States legally, documenting performance in a consistent manner, administering discipline in a fair and objective way, and terminating those who are not successful on the team without legal ramifications. Employment law itself is a vast and potentially treacherous territory, so someone on the management team must stay focused and up to date on changes, which are constant.

No matter what position has to be filled, and whether you decide to hire from the outside or promote from within, it all starts with a plan. What does the business need to carry it forward, what are the current gaps in personnel, and what skills are required for each position? Your first step is to think about the business and write down what it needs in terms of the people who make up the team.

Even a small veterinary practice needs someone with the specific responsibility of handling HR. This should be someone with appropriate skills and knowledge—or willing to be trained in those skills and acquire that knowledge—and it can be someone with other responsibilities as well. Some practices "outsource" this responsibility to HR consultants, but hiring, managing, and firing team members should not be taken lightly.

Most importantly, we want our practices to be places both where people like to bring their pets and team members want to come to work each day. The success of a practice is evident when people in the veterinary profession *want* to come to work for you!

CHAPTER 1: HIRING AND INTERVIEWING

The cycle of human resources begins even before people are hired to do certain jobs in your veterinary practice. That may seem obvious, but it can be difficult to attract and end up with the best person for the job. There are many steps to take that lead to a good hire, and the path can be long, especially if your team is understaffed at the moment. From advertising the position to conducting the phone interview and personal interview, to holding an observation day in the practice, multiplied by all the best candidates, this is a daunting task for any HR person. There is so much to learn about asking legal questions, obtaining the best information to make a good culture "fit," and following the process consistently for each applicant so you make an objective choice and don't just choose someone on a "hunch."

1

Where do I find or how do I hire qualified, professional support staff?

There are many resources to use when searching for potential candidates to fill your open position: newspapers, professional journals, websites, Facebook, Craigslist, schools, networking, and word-of-mouth. Consider advertising at local job fairs and going through an employment agency or your state unemployment office.

To ensure attracting appropriate candidates, never advertise without first assessing your needs and having an accurate job description. After all, you need to know what you are looking for. Conduct or review a job analysis to make certain you are considering the critical skills needed for the job and that the job description you are using is current. That will weed out many applicants who don't have the skills or experience that your practice, or the open position, requires.

Recruiting can be both an internal and an external tactic. Internal recruiting includes asking current employees to participate in job bidding (where an employee expresses an interest in and bids for a position), nominations (employees recommend people they feel would be excellent for the job), or succession planning (the practice identifies employees with high potential for a position and maps out a career path for their training). Practices can also regularly assess their teams for members they feel can be developed for future positions.

Consider posting the opening at your practice and in your practice newsletter. Your own team members or a client may have good candidates in their circles of friends, and merely mentioning or announcing the opening will elicit a response.

External recruiting is looking outside the practice for potential candidates. Candidates may be actively seeking a job, or not officially in the market but attracted by your ad (especially your reputation and brand image). Talk to other practice managers in your area to see if anyone has received a letter of interest from someone looking for employment in the veterinary field. Another manager may not have a position opening up and can pass along the information so you may contact the individual. (If *you* are a manager with a person's résumé in hand and another manager asks for potential referrals, always call the candidate first and ask if you may pass along contact information to other managers who are looking to hire.)

Finding the perfect fit is not always easy, but taking advantage of so many different options will ensure that you reach as many potential candidates as possible.

Do It Now

Take some time to review your previous hiring routines and see if a pattern emerges in where and what methods resulted in the best hires. This may help you concentrate your efforts and your budget, and contribute to the wording of the ad.

How do I effectively attract top-tier applicants when advertising for help wanted?

Top-tier applicants are those who have training and, ideally, experience and who envision a future in veterinary medicine. They are looking for not just a job, but a career. Whether that career is in your practice, or your practice is only a stepping stone, you will benefit more from this type of candidate in the time you are together. This type of employee is also already experienced, and perhaps even credentialed, in an area of the practice. To make it easier to attract and identify the top-tier candidates, include very specific requirements in your employment advertising. Here are some examples of how to present your needs:

1. You want to hire someone to grow the dentistry side of the practice; you'll advertise and only accept a Veterinary Technician Specialist (VTS) in dentistry or a credentialed technician with 4 years' experience in animal dentistry.

2. You need a solid technician who can integrate quickly into the team; you'll advertise for a credentialed technician with 3 years' experience.

3. You need a dedicated veterinary assistant; with the emergence of certification for assistants, this factor can be the filter for these team members. You also may want to ask for 3 years' experience.

4. You need a practice or hospital manager who can lead the practice into the future; you'll advertise for a Certified Veterinary Practice Manager (CVPM) with 5 years' experience.

It should be noted that your practice needs to decide how you define "technician," because there was a time when credentialing was not required in the profession, so asking for formal credentials may block qualified applicants. Do you want to hear only from credentialed (and perhaps not experienced) technicians, or miss out

5

on potential employees who have spent 10, 15, or 20 years on the floor but haven't obtained credentials? This is not a choice to take lightly, because one side will lose—the experienced-with-no-initials person may consider it a demotion to be called "assistant," while the credentialed technician will not appreciate the title of "technician" being used for someone who did not obtain the initials.

It is quite tempting to look at the stack of résumés and applications, and add candidates who are very close to your ideal but don't really have the requirements you advertised for to the "keep" stack. Remember that you want top-tier people, and only the best candidates will do!

Hiring staff based on nothing more than a résumé and first impression can be challenging. How do I ensure a good hire?

The best way to make a good hire is not to rely on just a résumé and first impression; many other pieces will help you make the right decision.

Always start by determining exactly what position needs to be filled. In other words, do not hire an assistant when the practice really needs a credentialed technician. Once the correct position is determined, it is time to get out that job description and see if it is still current. Job descriptions should be reviewed annually, and again specifically when hiring is about to take place. Let the team review the current job description and suggest any adjustments that might be needed, so you hire the person you need to do the job you need done. Then advertise for that specific position opening, using the minimum requirements and considerations in that job description. This makes it very clear when someone doesn't have the minimum experience or education required and should not even be considered for the position.

It's great to get an appropriate résumé from a candidate, especially if it comes with a well-written cover letter—sometimes that will give someone a place in the "good" stack—but you also need every candidate to complete an employment application. Some questions on the application will not be answered by a résumé. Those whose résumés demonstrate that they meet the minimum requirements for the job should be asked to complete an application.

The next step is the phone interview. A few basic questions will help you determine whether a candidate seems to have the needed verbal and written skills, disposition, and overall personality you seek. In that case, schedule an in-person interview. (By the way, when you catch someone on the phone, be sure to ask if it is a good time to call; be considerate of someone's possible location when they need to answer questions, such as the current place of employment!) If you aren't sure whether you've been given a home versus business phone number, schedule an interview via email.

In the interview, ask the same questions of each candidate for that position. This way, you can properly compare them to each other to find the best fit. Asking everyone the same questions protects against accusations of discrimination by ensuring fair and consistent treatment of each candidate.

The first impression is not the only thing to monitor during this meeting; dig deeper to find out what's under that good first impression. Use questions about how candidates handled different situations in the past, and see if their responses seem to match the practice culture.

If all goes well, schedule an observation interview. Ask the best candidates to come into the practice to see the flow of events, get a feel for the people and atmosphere, and demonstrate their skills and fit with the culture of the practice. This will serve two purposes: (1) It lets the candidate consider any need for additional information about the responsibilities and whether to accept or decline the position, and (2) it gives you a chance to see how a candidate might fit into the culture of the practice and the team a chance to weigh in. You will not be making the hiring decision alone; instead, it will be a team collaboration where everyone shares the responsibility for that hire.

If a candidate asks about being paid for a "working interview," it is wise to check with legal counsel regarding the law about paying for work performed before someone is formally hired. There may be liability concerns, among other things.

How can I conduct more effective interviews?

Many of us are not well versed in interviewing tactics. Take some time to educate yourself and brush up on your techniques before you start interviewing candidates. Keep in mind that there are some legal ramifications to conducting an interview.

Careful preparation is the best basis for conducting effective interviews. Prepare yourself, prepare your questions, prepare your documentation, and prepare to do more listening than talking.

What constitutes preparation? Reviewing the job description, using an appropriate advertisement, developing good questions, assessing responses carefully, giving information about the job and benefits, and listening to what the candidate is saying.

Useful questions are what are known as open-ended and behavior-based, rather than ones that just require a "yes" or "no" answer. Good listening skills (and some note-taking) will guide you from one question to the next and will help you get through any "rehearsed" responses a candidate may have prepared. The job description for the position will ensure that candidates receive an accurate picture of what they are getting themselves into. Finish up with a discussion of benefits and when the candidate can expect further contact.

Interviewing styles can vary between structured, unstructured, or even the group approach.

Typically, structured interviewing is specific to certain skills essential to the job. Structured interviews make comparing different candidates easier, because the questions are similar for each candidate. The unstructured style puts more control in the hands of the applicant by using interview questions that are more open-ended and may yield more information, and can change as experiences and skills are disclosed. Unstructured interviewing, however, may make it more difficult to compare applicants. Group, or panel, interviewing involves two or more people talking with the candidate. Group interviewing can be a benefit to the business, by getting active involvement of team managers in the interviewing process (for example, hospital administrator, tech manager, and lead tech interview a tech applicant). Group interviewing can also help develop interviewing skills in less-experienced employees by letting them watch and take part in the process (and may be deemed a succession planning technique for developing future assistant managers/managers). Regardless of the technique, preparation is the key to interviewing success.

Your final step for conducting an effective interview is to fill out an evaluation form for each interview, using it for making notes about skills or experience related to the job, which will help when determining whether a candidate is eligible for a second interview. Your best plan for effective interviews is to have a plan. Your interview plan should be part of the overall plan for the business—familiarize yourself with the needs of the business to identify the most effective techniques that work for your particular business.

What can I ask and not ask in a new-hire interview?

There are legal guidelines for what you can and can't ask in a job interview. Some relate to a person's private life and others relate to a work history. Here are some questions you should *not* ask, according to the Society for Human Resource Management (SHRM):

- Are you a U.S. citizen?
- Were you born here?
- Where are you from?
- What is your ethnic heritage?
- What is that accent you have?
- How old are you?
- When were you born?
- Are you married?
- Do you have any children? What are your child care arrangements?
- When did you graduate from high school?
- What church do you go to?
- What clubs or organizations do you belong to?
- Have you ever filed a worker's compensation claim?
- What disabilities do you have? (or Do you have any disabilities?)
- Do you have AIDS or are you HIV-positive? Do you have any other medical condition?

Here are some open-ended, probing questions that are recommended by SHRM and could generate some useful information about the candidate:

- Could you please tell me more about … ?
- I'm not quite sure I understood. Could you tell me more about that?
- I'm not certain what you mean by … Could you give me some examples?

11

- Could you tell me more about your thinking on that?
- You mentioned ... Could you tell me more about that? What stands out in your mind about that?
- This is what I thought I heard ... Did I understand you correctly?
- So what I hear you saying is ...
- Can you give me an example of ... ?
- What makes you feel that way?
- You just told me about ... I'd also like to know about ...
- Recommended reflection questions include the following:
  - So, let me say back to you what I thought I heard you say ...
  - So, that made you think (or feel) ... ?
  - So, you mean that ... ?

There are numerous professional websites that can give you ideas for questions regarding competencies such as decision-making, flexibility, initiative, leadership, teamwork, or time management. Explore these sites before you even place your classified ad, because they will help you with developing the list of skills and traits you want to see in your next new hire, what the ad can highlight, and what questions you want to ask.

RESOURCE

Society for Human Resource Management (SHRM). http://www.shrm.org. Contains a wide range of resources for hiring, managing, and firing.

ⅡⅡ➡ **Do It Now**

Put together an interview packet that contains the job description, application form, interview questions, interview plan for phone and in-person interviews, guidelines for an observation period, a tracking form to keep up with all candidates and where they are in the process, document dates and signatures, and an evaluation form. (It is a good idea to have legal counsel review your recruiting/interview process.)

What interview questions or other hints will help in selecting forward-thinking potential employees and separate the mediocre from the potential stars?

A mediocre employee is one who comes to work most of the time and does only what she or he is asked to do. A forward-thinking employee is one who is always looking to the horizon to see what is new in veterinary medicine, and how to do more with his or her position in the practice. To determine which type of employee sits in front of you as a candidate, try using these questions:

- What do you know about our practice? (The best candidates will have taken the initiative to research your practice, at least online.)
- Why would you like to work here?
- What is your understanding of teamwork? Discuss some teams you have been involved with—both a good team and an ineffective one. (The past is the biggest predictor of the future.)
- Describe a situation where you planned and implemented a change in the practice.
- Describe a challenging incident/conflict from a previous work experience that has facilitated personal growth. How will it help you here? (You want someone who is not necessarily perfect, but learns from his or her mistakes and moves forward.)
- Describe a situation in which you had to use your presentation skills to influence someone's opinion or decision. (Each employee must be an ambassador for change to create a new and better future.)
- Describe a time when you were faced with a stressful situation that required your coping skills. (It is best if someone knows

what steps to take before encountering such a situation in your practice.)

- What are your professional goals? Where do you see yourself in 1 year, 5 years, 10 years? (Forward-thinking candidates already have a good idea of how they want to steer their lives and their careers.)

An in-person interview can be complicated by many factors, such as manager availability, candidate mood, and limited time to talk. To learn more about the candidate, you could provide a list of questions for candidates to take home, answer in their own "space," and return to the practice for consideration. If a candidate doesn't complete and return this document, he or she is bumped down to the bottom of the heap. You want an employee who will go above and beyond—someone who won't do that during the recruiting process is likely not to do it once on the team!

How do we find out if a potential candidate will fit into our culture?

The telephone interview is a starting point for narrowing the field of applicants. It is a great way to assess whether a person has the skills and qualifications you are looking for, and a good way to assess communication skills.

The site visit interview is an even more effective way to assess whether someone will fit into a practice culture, because there is only so much you can learn about a person on paper or through testing. The visit can be handled in a number of ways, from one-on-one to a panel. If the person is going to work in different areas of the hospital, having a few people involved in the interview process might help with assessing whether the candidate will fit. These people should be well versed in the interview process and have specific areas they will ask the applicant about.

A word of caution: Some places like to do a "working interview" where they leave the candidate in a room full of "regular" team members. However, this is when the other team members may put the business at risk by asking questions that should not be asked during the interview process. Be sure everyone involved in interviewing potential new hires is aware of what they may and may not ask.

Skills/aptitude tests and personality tests are also useful tools. Just make certain you are not violating any rules or laws when it comes to these pre-employment testing processes. You may want to have them reviewed by a professional for reliability, validity, and non-discriminatory methods.

Always have a business plan for the skills you want the interviewee to demonstrate. You may want to have a pre-arranged "practicum" as part of your interview process, instead of merely hoping that relevant cases show up while the candidate is in the middle of a working interview. The candidate may successfully advance through the phone and in-person interviews and become part of the finalist group to demonstrate skills you have deemed essential to the job function. If the next round of interviewing is a practicum (not a working interview), check with counsel regarding validity and non-discriminatory approaches.

Before even starting your interview, know what the business needs, know what questions you want answered, know what skills are essential—in other words, do your homework first and set up your plan for this particular interviewing session. In a few months, if you need to interview again, review your plan and make adjustments based on the changing needs of the practice, the position you are interviewing for now, and the skills required for that position.

Ⅲ➡ **Do It Now**

Select or create 10 interview questions to get started. Consult with your attorney to make sure your interviewing material or plan does not create any legal problems.

How do I select a good office manager who can deal with doctors and staff?

The right choice begins with the needs of the business. Before you even place the ad or scan your current team for that special person, you must have an honest commitment as the owner of a business to focus on what you really need. This will lead you to the next steps of creating or updating a job description, developing behavior-based interview questions, and placing the ad and opening the position up for inside people to apply.

There is no cookie-cutter standard set of traits for every veterinary hospital's office manager. That great manager at the practice down the road may fail at your place because of a different culture. There are some qualities that everyone will list—organized, dependable, upbeat, hard worker, flexible, personable, with certain computer skills essential to running the business. Still, these are very general— and can mean different things from one practice to another.

When looking for an office manager, consider succession planning. This involves looking at the entire organization and the leadership potential of current employees and potential new hires, not just about the next owner. Effective succession planning involves defining leadership criteria, having a plan for developing these criteria, and monitoring the development of individuals into certain roles and responsibilities. The office manager is often a key element in succession planning—that's the person who tends to know essential details, from computer systems to client names. The right office manager may help during times of turnover and can assist a practice with continuity of care and in promoting the mission and vision of the practice.

To help with crafting a job description, consider asking your team what they need from an office manager. If this office manager is to manage the doctors in any way, ensure that the scope of authority is included.

It is often true that "common sense" is not that common. While you may expect professional behavior from a new team member who is excited about learning, you might get someone who has never had to *be* professional. You also need to keep the training program energized and interactive to really catch the attention of the new hire.

The training program is paramount when it comes to producing good team members. New team members have to be trained on how the practice expects them to perform, and existing team members need to continue to improve their performance and learn new things. Implementing phase training and establishing levels of competency will encourage continued advancement. Monitoring the training of each employee is essential. Along the way, there will be those who are promoted to leadership roles, and those who are dismissed after a short period because they are not a good fit. The manager must stay focused on the success of the business while considering all aspects of building great teams.

How do I teach people what it means to behave and act in a professional manner?

A team that understands professionalism is created one step at a time, with each person who is hired for the team. It helps to have standards in place to apply to all employees.

During the search process, look for professionalism at the starting gate by paying attention to the quality of the information in a résumé or application; inclusion of a cover letter that is well written; an answering machine that has a professional yet friendly outgoing message; pleasant and appropriate discussion during the phone interview; showing up on time for the in-person interview; dressing properly for the interview (and the observation period, if offered); and behavior around other team members and clients during the stay. If candidates do not behave professionally while being considered, don't bring them on board.

However, even people who present themselves professionally at first may let those standards decline once they have worked in a practice for a while. When this happens, it's time to take a good look at ourselves in the mirror. Did you look for or ignore signs from the hiring phase? Did the job description spell out the type of behavior that is expected? During training, did this new member receive instruction on not only what to do, but how to do it, specific to communication and behavior when working with other team members or clients? Perhaps most importantly, are you and your current team serving as effective role models for professionalism?

There are practical action steps to ensure that you have a process in place that produces quality team members who know what level of professional behavior is expected at your practice, including job description, leading by example, and rewarding appropriate behavior. Professionalism is not a "teach once and forget about it" topic.

Everyone slips up now and then in professional behavior. To mitigate the risk of slipping during a crucial relationship (with clients, a referral practice, or your own team members), make the topic of professionalism part of the culture of your practice.

In the job description for each position, include the "soft skills" that are required for employment and retention in the practice. Define these skills so everyone understands what behavior is expected. The best way is to get the team involved in these definitions. Put "professionalism" in the job description, then list the behaviors that constitute professionalism: wearing a uniform or following a specific dress code; keeping the uniform neat and clean; consistently wearing a name tag; maintaining a comfortable level of eye contact; smiling and greeting both the team and clients appropriately; and refraining from personal activities in view of clients (eating, using a cell phone, talking loudly—and negatively—to other team members, focusing on a computer screen rather than speaking to a client, handling emotions well by staying firm yet welcoming during negative encounters, etc.). Professional behavior can be taught using role-playing and sample scripts, or dissecting an adverse encounter to learn from the event.

You are still not done; as a leader in the practice, you must also lead by example. Team members will look to you to set the tone of the workplace and work activity. You must also demonstrate respect to each team member if you hope to get it in return. Make professionalism a goal, teach the team to behave in this manner, and set a good example for the team.

Observe and reward professionalism by praising a professional action at every team meeting—read a client comment, give someone a pat on the back, tell a story. Let the team take part in this recognition by sharing their observations of professional behavior in the workplace. Professionalism will become a way of life in your practice.

10

How do we make team members who have been together for a long time realize they still need to maintain professionalism, even though they are friends outside the workplace?

We expect our teams to be professional when at work (see Question 9). We expect personal lives and personal issues to be left at the door when employees come to work. We all feel that home time is the employee's time and should not be controlled by the business. But the truth is that the line between friends and co-workers, and even between home and the workplace, often becomes one big blur. We may love the idea that we have a "family" business—but how many managers feel like the parent of 30 kids?

Sometimes the most complicated workplace friendships are the ones between managers and employees. How many managers have Facebook friends who are employees? How many managers want to reprimand employees who call in sick and later post on Facebook about what a great time they had at the park? The lines can be blurred, and legal pitfalls abound.

How often have owners' or managers' friendships with employees brought up the question of fairness? Often, doing what is best for the business means making hard decisions and ignoring the friend relationship. Just be prepared: Repercussions will occur, there will be fall out and friendships will change.

No matter who the friends are, a lot depends on the maturity of the individuals involved and the culture of the business. In many businesses, family members and friends all work together productively and healthily, while other businesses suffer with the same set-up.

The take-home message is to keep it professional, manage expectations, and discuss areas where a line is being crossed. You are not always going to be prepared for the unprofessional workplace behavior of an employee, the Facebook rant, the barroom fight between co-workers, or the divorce of co-workers, but professional and consistent responses will show the team what is expected and what the consequences of failing to maintain respect and professionalism can lead to.

11

How do I encourage the staff to take advantage of the many learning resources that are now available (online, etc.)?

We have all been there—announcing conferences, posting signs about online webinars, talking about pharmaceutical company online classes. Yet we don't see enough team members take advantage of these opportunities.

A reward is a powerful incentive. Make it clear that the practice will pay or reimburse for all or part of the costs of continuing education (CE), including attendance at out-of-town conferences. Offer raises, promotions, and even restaurant or entertainment vouchers for successful CE efforts. Celebrate every time a team member passes a course or obtains a certification.

How many of us set out with a strategic plan for the year in terms of conferences? Why let CE attendance be just about who is speaking and where the event takes place? Today's businesses must survive in a changing economy. If the business is conducting a yearly strategic planning session, why not take a look at the team and the skills or knowledge you will need team members to have in the upcoming year?

Set up a CE budget for the year for team members. Sometimes we let the conference be the driver when the business needs to drive the team through learning and CE opportunities. You assess your team's performance, so why not help them grow by becoming more involved in helping them attend specific sessions? You may have to do the "leg work" of signing them up and providing the space and time to take an online session, but the outcome is a win–win for all involved.

For those who do attend CE meetings, talk to them at the time of registration. Go over the strategic plan with your team member and ask them to attend some sessions specific to the needs of the business. Ask for a simple report—one idea to improve business functions, one idea to improve the client experience, one idea to improve the care the pets are receiving, and one insight about a favorite session (they may be totally unrelated to your business, but what fun to hear about llama care or bird diseases!).

Equally important to planning what CE to attend, is planning how to implement ideas from CE attendance. There are few things more discouraging to your team than ignoring ideas they bring back from a meeting. This step requires buy-in from the owner, a thorough understanding of the practice's vision and goals, and the ability to set up a project team to work on implementing an idea. Perhaps a CE session on a new procedure is available and your practice's Strengths, Weaknesses/Limitations, Opportunities, and Threats (SWOT) analysis designated implementing this new procedure as a goal for the year. Tell your team about the goal and set up a project team—who will attend CE, who will work on evaluating the information and developing the project, who will teach the team the new information, etc. If the team knows about the relevant goal of the business and that you value their contributions, they will see the value in attending CE and participating on a project development team.

Ⅲ➡ **Do It Now**

Perform a SWOT analysis of your training program to identify Strengths, Weaknesses/Limitations, Opportunities, and Threats and see where current employees might be plugged in to advance and strengthen the team.

12

How do we create, present, and track training for new and existing staff?

Do not leave training up to chance. This will only lead to frustration for the newly hired person and the rest of the team in dealing with the effects of an incomplete training process. Develop a training checklist and require documentation from employees who take training.

The checklist should be more than just a laundry list of tasks; it should also schedule the dates of training and assigned coach/ trainer, whether a local person or a current employee, to each topic. Train the trainer first (regarding tasks to teach, teaching methods to use, documentation, and assessment) and then let the trainer take on the task of training a new hire.

In addition to being of legal benefit, training checklists can help management guide an employee in personal growth and expansion of skills and knowledge. This goes beyond the initial hire date and well into the future business relationship.

Many systems are available to electronically track training that a person has received. Although you may not have to go into as much detail as some programs provide, keeping a tracking sheet on all your team members will prove invaluable as the practice grows and changes.

Training checklists aren't just about training someone in a new skill. They are also about HR issues such as safety, sexual harassment, management, and other issues pertinent to the overall business functions and legal responsibilities. Certain subjects require yearly training sessions and documentation of who attended. At the start of every year, schedule your topics, identify who will conduct each session, and post the information where every employee can see it—on a bulletin board, at the website, in pay envelopes. File the material presented and the signed attendee list for legal proof of having provided the training as required.

Lest we forget, face-to-face conversations are the backbone of building a relationship, and these interactions are vitally important during any training sessions—but especially important for a new hire. Take time (even if it is once a week) to meet with new hires and discuss everything they have witnessed, skills they are being trained on, and any areas of confusion they may be experiencing.

Do It Now

Create or update a list of training topics to offer team members for the year, based on current staffing needs.

13

How do I implement phase training?

Most phase training involves lists—lists of duties and timeframes to learn and perform those duties. Phase training is useful to a practice in two areas: training a new hire, and developing a current employee who is looking to take on more responsibilities or work in a different area of the hospital.

Any training should begin before a person is assigned a training date. Training begins with the manager conducting an analysis of what skills the practice needs to best serve its clients and patients (needs assessment), then identifying the resources for providing the training, and finally designing and developing the appropriate training sessions. Once these initial steps are complete, the manager can make assignments and implement the training. Afterward, an evaluation of the training session and knowledge acquired must occur.

A training program should have an objective and a goal in mind from the beginning. Failure may result if you haphazardly throw someone a laundry list of tasks and tell them to learn what they can.

Suppose you analyzed a new hire and have determined you want to train that person in how to handle exam room duties before they receive surgery room training. There is a specific objective and goal, so you develop the training program and checklists with those in mind. Determine the amount of time needed to learn and, once the training is finished, conduct an evaluation of success so trainees know what they should learn by what date and how they will be assessed.

Another type of training is based on analysis of the job. Your practice has decided to add a new service, so the manager assesses what skills and knowledge will be required by the team to perform this new service. You design, develop, implement, and evaluate training for this specific service.

Any training has a chance of failure. Keep your eyes open for common reasons, such as lack of direction, objectives, or accountability; or failure to allocate sufficient time, prepare the trainer to train, or consider all the ways adults learn.

Remember, phase training isn't just for a new hire. It is based on the strategic planning of the organization for the future needs of the practice. It should involve both new hires and new needs of the practice.

Ⅲ➡ **Do It Now**

Look into resources for phase and general training from AAHA and local resources.

14

How do I adequately mentor new employees?

Mentoring involves building relationships. It is about the strategic direction of the practice and team members growing in their responsibilities. A mentoring relationship matches a new team member to a more experienced co-worker to help the new person gain skills and knowledge through observation and one-on-one interaction. Mentoring can occur at any time during employment and is between another professional (sometimes outside the practice) and an employee who is either new to the practice or seeking to learn a new skill.

A mentoring program should involve setting formal meeting times, goals, and reviews. Often, a mentee requests a specific mentor (someone in a position the mentee aspires to). A mentor is a teacher and a facilitator. A mentor should challenge the mentee to move beyond the comfort zone by providing safe learning situations for taking risks.

There are other methods of developing employees that are similar to mentoring. Coaching can occur between a manager and an employee, and focuses on developing a person in a current job. Coaching is about function and tasks performed at the team member's current level, while mentoring is more about professional development that can take employees beyond their current tasks. Coaching involves evaluations of an employee's current job duties and performance to help them do those duties well or better.

Buddy systems are similar to mentoring but more on the level of peer-to-peer learning, with employees being matched with peers in the same work departments or with the same titles to "mirror" and encourage each other as needed. These match-ups are usually more to help a new employee adjust to the job than to help someone advance in the practice.

When determining a program for the new hire or a seasoned team member, consider whether they need a buddy, a coach, or a mentor. What is the team member's career goal and what does the practice say it needs from the team member? Using input from the individual and from the practice, you can create systems to encourage growth via the help of a buddy, a coach, or a mentor.

Be sure to keep in mind that mentoring is usually a volunteer effort. Recognize and reward team members who are willing to give that extra amount of time and effort to a colleague.

⫸ **Do It Now**

Establish systems to handle buddy, coach, and mentor relationships by defining the purpose and goal of each system, evaluating current team members who would be excellent buddies/coaches/mentors, then working with those team members to evaluate people needing attention (i.e., needing one of those systems to acclimate/improve or develop potential).

15

How do I get senior staff to set a better company example?

For senior team members to be good role models for the other team members, they need to understand and feel their importance to the practice. They will know, better than anyone else, the influence and value that come with seniority. Value their opinion and seek advice from these team members. Establish this pattern with the rest of the team—when someone comes to you with a question, send them to the senior team member who can help the most. Let these senior team members know how much you count on their contributions, and that they will be a source of information for others; incorporate this role into job descriptions. Tell them how much you appreciate their being a "go to" person for the team, and how valuable they are to the practice and the team.

When it comes to assessing "soft skills" in performance reviews, ensure that the more senior people become, the more they are expected to set this good example. Hold them accountable for being someone to rely on. Even before reviews, support this concept of seniority in the job description and training phase.

Let senior team members take an active role not just in training, but in mentoring others (see Question 14). Match the personality or the specific area or skill, and formulate an official but casual process to monitor the progress of both the newer team member and the senior one. Each has a responsibility to meet the goals of professionalism and team efforts (see Question 9).

16

What is the best way to encourage and train for proper telephone etiquette?

We all know the saying about the power of first impressions. Since the telephone is where prospective clients usually get their first impressions of our practices, proper phone etiquette is very important to the image of a practice, so training in that task is essential. Research has found that 99 percent of clients have contacted the front office via the practice phone number before their arrival.

The front desk or receptionist has to handle a wide variety of calls, so training must include lessons in all of these types:

- New client
- Existing client
- Veterinarian from the area
- Member of the support team from the area
- Colleagues
- Family members and friends of colleagues

The basic etiquette for phone calls is simple courtesy. You might want to prepare a script for answering the phone and distribute it—not just to the front desk, but throughout the practice, because different people may answer the phones at different times. You want the phone to be answered with "Hello, this is the XYZ Veterinary Clinic. How may I help you?" rather than, "Hi" or "Yeah?"

One thing to strive for during all phone calls is a personal connection. It's been said that 40 percent of body language or non-verbal communication is based on tone of voice. This is the most important tool when answering the phone. It is possible for a client to "hear" a smile (or lack of one) from the person answering the phone. You might want to train receptionists by positioning a mirror at each phone station as a reminder to smile as they answer the phone.

Suggest that people take a couple of deep breaths before answering, because that can help busy team members focus on calls, as well as control their voices and expressions. Remind everyone that a given call could be the most important task for that client at that moment, even if team members have answered a hundred calls that day. The next client should not have to suffer from a brusque response just for being call #101.

Useful training methods include role-playing or acting out different scenarios. The team may groan when you announce such training, so be sure to have some fun with it. People will remember things that involve a good experience more than an activity that is dull or by rote.

It helps to record some of these calls for playback and discussion. Listening to the conversations together can be an educational tool and provides an opportunity to do better. For long-term team members, use these recordings as a measure of their performance. If you are going to score or rate them on performance, let them hear exactly why they are receiving that score.

Another good training tool is simply opening up the employee's awareness. Have your front office team members document each opportunity for a phone contact for a week, and rank each interaction on a scale of pleasant to disastrous. That will help people gain an acute awareness of how it feels on the other side of the desk, which can be easy to forget in the daily hustle.

17

How do you get the client-relations team to be more confident in answering client questions and educating clients, and to feel important?

Confidence is something that you, as a manager or team leader, can help an employee build. It is also something that you, as manager or leader, can destroy. Once someone's confidence is undermined, it is not easy to get it back.

If clients sense a lack of confidence in a service team member, they will disregard any and everything else that team member has to say. An indication that this is a problem is when you hear, "I want to talk to the manager ... the doctor ... the technician ... the owner." It isn't fair in most cases, but a lack of confidence can be taken to mean a lack of knowledge, understanding, or connection that means the client isn't getting the answers they need or want.

First and foremost, find out what questions are causing this discomfort for the front-office team. This could be medical terminology, commonly used medications and their uses, lab tests and what they indicate, etc.

Don't forget to also focus on pronunciation! The veterinary world, like any medical profession, uses a lot of words that people don't automatically know how to pronounce. If an employee utters a word that doesn't come off the tongue easily, stutters it out, or hesitates or acts embarrassed, that is like a drop of blood to a shark, and that client may proceed to tear you up!

The front-office team—in fact, every team member—should be encouraged to ask questions, learn more about what they don't know, and understand the importance of never making up an answer. Tell them to be confident in their not-knowing—that it's better to ask for help than appear flustered or uninformed, and that it's perfectly OK to say, "I don't know." Provide a script—if a client asks a question that the team member doesn't know, they should respond with confidence, "That is a really good question, and I don't know the answer ... but I'll find out and let you know!" Every time one of the team comes to you with a question or subject of concern, or even makes a mistake, be gentle—don't attack or call someone stupid for not knowing an answer. Encourage initiative, but make it clear that bluffing is not acceptable.

Once you identify any problem areas, build a plan to train or educate the team up front so they will feel more confident. Consider holding weekly sessions where team members can let you know the kinds of questions or requests give them trouble, and provide the information they need to handle those matters better the next time around.

When a client shuts down with a front-office team member and asks for someone else to talk to, it is easy for that team member to feel unimportant, untrained, and in the wrong place. The truth is, some clients will always want their question or concern taken one step up the food chain, to show that their concern is valid. This should not be taken personally, but it's hard not to. Let the team know that you understand.

The front-office team will feel more important when their role on the health-care team becomes more important, and that comes with education and reinforcement. Be sure the rest of the team appreciates the front office team, since every person who calls and every animal that comes through the door has the potential to be converted into a client and patient during their initial interaction with the client service team.

RESOURCE

WISC-ONLINE. http://www.wisc-online.com/Objects/ViewObject .aspx?ID=gen504. Provides audible pronunciation of medical terminology in different categories, such as body structure, cardiovascular system, digestive system, integumentary system, and more.

18

How much time do you give to the new employee you thought was the right candidate but who does not show competency during the training process?

As little time as possible! While that may seem like a flippant answer, there is truth in it. Even if you go through a stellar recruiting plan, an intensive interview process, and all the other steps, from calling references to conducting tests, you can still end up with someone who doesn't fit in or has misrepresented his or her skill and/or knowledge.

That's about standard: After a few weeks, you should have a sense of whether a new person is fitting in. Even if someone has fantastic skills and a solid base of knowledge, the team is not going to give the person a chance to show off unless they feel that the employee fits into the culture. This is another reason it is important to involve the current team in developing training lists, buddy systems, and other methods of incorporating a new employee into the team. It becomes a group effort, and everyone has a responsibility for giving that new person the opportunity to succeed.

If it is clear very quickly that a new person's skills and knowledge are an issue, counsel the new employee, review the training plan, see if another coach or trainer would fit this person better, try delivering the training program itself in a different way to respond to that person's individual learning ability, and be sure it isn't "you" or "the practice" that is the problem.

Whenever there is an issue with a team member, whether someone has been there 10 years or 10 days, it is time to look at yourself in a mirror, and the practice with a critical eye, to make sure you have explained expectations (in other words, provided a job description), implemented training so the employee can reach those expectations, evaluated performance fairly and honestly (through a performance evaluation or assessment), and provided sufficient support and encouragement.

Typically, if someone is a good fit in terms of personality and work ethic, but the problem is skills and knowledge, everyone will work harder to determine if this person is "teachable." If team members demonstrate that they can be taught, more time or different tactics may be needed to bring them up to speed. This will be made possible by the team accepting the person into the fold.

19

How can we get employees to pay more attention to errors and missed fees?

One way to get employees to pay more attention to errors and missed fees is to show them how much money they're throwing away. One source estimates that the average practice loses 15 to 25 percent of its income due to uncaptured charges—unpaid bills, charges not made, waste, etc. Tell the team that these amounts could have been their raise, bonus, holiday party, new equipment, etc.

To begin, assess how many charges are being missed (and if there are any trends) by performing routine, consistent audits of patient medical records. This takes time, but it's likely that the wages for one person assigned to do this with every invoice will still be less than the cost of lost charges. This would only be a good investment in an extra team member if the charges are found before the client settles the invoice. Otherwise, if an audit takes place after the visit, it identifies the problem after it is too late to charge the client. Second, go to the team with the details of these missed charges to determine how it got missed. In what area of the practice are we missing the most charges? Is there a way to change protocol to make misses less likely?

To bring these ideas to life, you could make a game with chart audits to see who can find the highest/lowest missed amounts (smaller teams sit with 10 charts and go through them together). When mistakes are found, add these to a chart that tracks certain information: date of missed charge, patient first/last name, item(s) that was missing, how much the client would have paid for those missed items, and finally the total of all the missed charges for a set period. The total will likely surprise the team! Third, discuss the effects of missed charges and the importance of capturing all services and products. Talk about how the practice could have invested that money for the team or the facility.

Once employees understand the importance of capturing all charges, there are plenty of steps to take in addressing errors and missed charges. One is to make it easier to capture costs by standardizing billing. At each point of the visit, the patient's charge sheet should be checked for completeness; multiple sets of eyes are better than just one. The topic can be discussed at team meetings, with discussion about any particular item that seems to be giving the team trouble.

When an employee is found to have made a mistake or omission, pull the person aside to review what went wrong and be sure they understand how and why to charge for that item. A practice may decide that, with management's approval, an employee who misses a charge must call the client to ask for further payment. That will probably be so uncomfortable that the employee will be more motivated to do it right the first time, rather than have to repeat that embarrassing experience. If it happens again, that team member can be reprimanded through warnings and, if necessary, eventual termination. Just make sure that you have a written guideline on what constitutes an error and how many times you will let someone commit it before termination is likely—and that you record every instance of noncompliance.

Ⅲ➡ **Do It Now**

Create a chart illustrating missed charges and income to start showing the team what they're missing and hold a team meeting to discuss how to fix the problem.

How do you promote team members to roles that require more challenging tasks, responsibility, and accountability while setting them up for success?

The first step may sound simple, but it's one that is often forgotten: Ask if the person *wants* more-challenging tasks and the responsibility that goes with them. You may have selected someone for a new role because they were the best at their current job, but not everyone wants to change jobs, do more or different kinds of work, or have more responsibilities.

Before a team member decides to take the new role, discuss the difficulties as well as the advantages to the move. If appropriate, make sure the team member knows that she or he will be spending less time on the floor, for instance, and more time on these new duties. Explain how the team dynamics may shift. Some team members may resent this person's new role, or that particular team member taking on that role. Close friendships could become strained if someone takes on more responsibility and now supervises people who used to be peers.

Try to cover the bases, because, once a team member has been promoted, it is difficult to step back down into a previous position—and you may lose a good employee over it. Consider providing management training for someone who is being asked to function as a supervisor, or technical training and certification for someone who wants to advance to that level.

In the best-case scenario, the team member has come asking for a more challenging role, so the first step may be done for you. Ensure that the team member has given some thought to this new role. Ask them to present a proposal for the move they want, outlining both advantages and disadvantages. This helps you to see if they are prepared enough, or whether more discussion is warranted before a decision is made. You don't want someone to have expectations that are unrealistic—either positive or negative.

During the transition in roles, "feed" the new tasks one "bite" at a time—give the team member a small task, and see how that goes by checking in often and discussing the outcome. As the team member demonstrates that she or he can handle more, more challenging tasks can be added. Meet with the team member on a consistent basis, perhaps more often at the beginning and then tapering off in frequency as the team member makes positive progress. If applicable, provide continuing education on the topic of their position, such as suggesting and providing relevant books, articles, and online or in-person classes.

21

How do you find and/or create leadership within the organization?

It is imperative to realize that there are leaders all around your practice—some you want and some you don't. If you abdicate your own leadership position and responsibilities, you may end up with a leader by default and an unwelcome culture.

Leadership is best developed in your team by building respect, trust, and responsibility. Respect the ideas of your team and show appreciation for the work they do. Encourage open-door communication between managers and team members. Train and coach people to develop their leadership potential, and give feedback to enable every employee to be a valuable member of the team.

Leadership training isn't just for owners and managers. You can build leadership through team projects and workplace groups. Although the owners are considered leaders by default, so are the associate doctors, head techs, or lead receptionists. They may not have managerial duties, but their positions put them in the forefront of leadership evaluation.

To identify potential leaders, pay attention to how employees interact with each other. Look for who seems to have earned the respect of colleagues, who shows initiative in taking responsibility or suggesting new ideas, who helps other team members and who goes the extra mile in getting things done.

Leadership training does not mean sending someone off to college for a degree. Rather, it means delegating authority, assigning tasks, and giving more responsibility. Think of all those projects you know you should be doing but don't have the time to do yourself—training, website management, social media coordinator, safety, coaching or mentoring, client education—and look to your team for those people with the skills, strengths, and leadership potential in those areas.

Be aware that leadership sometimes comes from the bottom up. Why not take an active role in mentoring others and developing an organizational culture where learning, initiative, and collaboration are emphasized?

Do It Now

Draw up a rough organizational chart to identify areas where stronger leadership might be needed and plug in names of team members who have demonstrated leadership potential.

22

How do you train managers to manage their staff?

When a practice is just starting out, it usually has a relatively small number of employees. The owner typically is also a practicing veterinarian who works hard to keep everybody on the right track, but might be paying more attention to hands-on animal care than day-to-day management of the practice and its people. As the business grows, the team grows, and the owner comes to realize that help is needed in managing the people.

First, you must train these managers. Most of the team members who move into management began as good front-office team members or fabulous technicians. However, success in those roles does not necessarily mean they understand how to manage people.

The person who is stepping into a management role has to know what they are facing. It's not as easy as thinking, "I know how to do that job, so I just need to watch others do the job and make sure it's done right." That might be the objective, but the process can be difficult.

You can prepare your employees to move up on the career ladder by letting them know what management actually involves—that they will now be spending some of their time on working to motivate team members; ensuring everyone is treated fairly; organizing the workload; helping settle conflicts; and handling performance issues such as coaching, counseling, discipline, and perhaps termination. People who only take a step up the ladder for an increase in pay will only be sustained by that move up to a point.

It is vital to train people on how to manage. Communication skills will be key; *how* we say something is as important as *what* we say. Initially, have a new manager run all communications through a management mentor. The mentor will check for tone, more than anything—whether this communication could offend or inflame the team.

Before every in-person meeting with their team, the new manager should be required to explain the situation to the mentor. The mentor will make sure there is an issue that should be handled by the supervisor. Then they will discuss word choice, body language, and attitudes that could affect the success or failure of the conversation. Over time, the new supervisor learns by being shown what is best. The mentor can start loosening the reins a bit as confidence increases for both of them.

This is most important in helping make the transition a smooth one for the whole team, and getting the new manager off to a good start.

As for more management tasks, provide training and confidence-boosting by sending the new manager to continuing education programs in person, over the Internet, and through books and journals. Debrief after each of these events, to be sure the new manager is taking away the proper points for implementing ideas for the practice.

RESOURCES

Learning Theories. http://www.learning-theories.com/addie-model.html. Contains information on learning theories, such as the Analysis, Design, Development, Implementation, and Evaluation (ADDIE) Model.

AVMA PLIT. http://www.avmaplit.com. Contains information on employment practices and liability insurance.

CHAPTER 3: COMPENSATION AND BENEFITS

People typically enter veterinary medicine from a desire to help animals, and are excited to get that opportunity by working in your practice. There may be times when these emotional rewards and recognition can be given without a price tag—sometimes these rewards are valued the most—but you also have to provide a competitive compensation and benefits package, appropriate to the industry and your local region, that allows team members to do their work in return for wages and benefits that are more than minimal. From determining starting pay, to how and when to deliver wage increases, to what type of benefits package to offer, there is a lot to consider.

How do we know if our compensation is comparable to other practices?

Each practice is its own little world, collecting revenue and paying expenses within its walls. When it comes to payroll (the largest slice of the practice revenue "pie"), we know what percentage of our revenue is used for wages, but this doesn't tell us whether we are competitive in what we pay our team members, especially the good ones. Being competitive is important in attracting and keeping good employees, so it's important to have a sense of where your practice fits in the larger world.

Look at employment ads to see what other practices are offering for various positions. You could check with local chapters of professional organizations to see if colleagues will share their compensation information in general terms.

Professional organizations measure compensation for general types of positions (in other words, not a veterinary receptionist, but a receptionist in general). The American Veterinary Medicine Association (AVMA) and the American Animal Hospital Association (AAHA) provide compensation and benefits measurements in a meaningful way, taking into consideration education, years of experience, years in a practice, and geography, among other factors. The Veterinary Hospital Managers Association (VHMA) generates a compensation survey every 2 to 3 years.

Compensation should be based on skills, training, and experience, and must comply with local, state, and federal guidelines for minimum pay. Team members must feel that your system of assigning wages is equitable, so clear communication to the team needs to include discussion of how wages are set in accordance with skills, knowledge, attitude required for the job, and performance displayed by the person (and subsequently discussed in performance reviews).

Initial wages should be equitable by rewarding previous education, experience, and training at the time of hire.

It's often a good idea to seek legal counsel in confirming federal and state minimum wages, which employees should receive salaries or be paid by the hour, and who is exempt from overtime pay.

Keep in mind that benefits are an important part of this conversation. Besides the hourly wage, what other benefits can be offered to a candidate or employee? If you pay a little less per hour than another practice, you might offer more-generous health insurance coverage or extra vacation days. Having a retirement package will attract people, as will paid holidays, discounts on services for employee pets, certain shifts that fit employee lifestyles—there are numerous possibilities for enhancing wages and benefits without costing the practice a fortune.

RESOURCES

American Animal Hospital Association (AAHA). Updated biannually. *Compensation and Benefits.* Lakewood: AAHA Press.

American Veterinary Medical Association (AVMA). Updated biannually. *AVMA Report on Veterinary Compensation.* Schaumburg: AVMA.

Veterinary Hospital Managers Association (VHMA). Updated biannually. *Compensation and Benefits Survey.* Alachua: VHMA.

⑾➡ **Do It Now**

Develop a total compensation worksheet to spell out all of the benefits you offer and demonstrate the total amount that is being invested in each employee (see Question 40).

When hiring an experienced new employee who expects a higher salary than current staff, do you have to raise the salaries of all staff so the new person's salary is in line?

The short answer is "no," but it's important to know why not. You must assess your current wage structure to ensure that it is fair to all employees, not just a potential candidate who is making a great impression in an interview.

These are many of the basic fundamentals to structuring wages:

- *Range of wages per position:* For each position in the practice, there should be a corresponding range of wages identifying the lowest and highest wage a person in this position can expect. This will be determined by the average ranges for the same position in other veterinary practices and similar fields in your region, what you have learned through the years about what it takes to hire and keep topnotch employees, federal and state wage guidelines, and benefits the practice offers that can enhance any wage that seems low.

- *Levels of positions:* Not every person in the same position will be comparable to everyone else in that position. This is especially true with technicians, where your top—Level 4, for those practices that use a formal level system—employee may be able to monitor anesthesia in surgery, but your middle—Level 2—technician cannot do that same task. These levels must be defined by the skills and knowledge comparable to the difficulty of the level and any special training or certification that might be earned by different employees. To move from one level to a higher one, an employee will have to demonstrate having acquired the skills and knowledge to be promoted.

- *Pre-existing factors:* A candidate's place in the practice may be determined by where they were previously employed, for how long, and at what type of practice. Other factors include acquired education and credentials.

Even with that data, it can be difficult in the hiring process to determine if someone's education, experience, and credentials will match your practice's compensation levels.

One method of setting the starting wage of a new person is to hire at a lower range, and increase their pay as they make their way through the practice's training system. Let's say that you hire a candidate as a technician whose experience and education would qualify them as a Level 2 technician, based on the information you have from the application and interviewing process. Yet, you offer a somewhat lower wage to come in and prove that their skills and knowledge base equate with a Level 2 technician in your practice. After all, that person has to start at the beginning of the training program to learn the basics about your facility. That might be considered an introductory period, with the wage going up after a stated time period and reaching stated milestones.

You may have found from past experience that an average employee can make it to a Level 2 technician within a matter of months. A new hire who advances accordingly shows they belong at that level. When they reach Level 3 requirements, increase their wage to where it should be, based on the range.

Most practices have a policy forbidding employees to discuss their wages with other employees, but we know that employees *will* talk about their compensation and benefits, and the National Labor Relations Board (NLRB) now says they have a legal right to do so. In the best interest of the practice, make certain that the following is true:

- Compensation calculations are known among employees.
- Regular performance reviews occur to discuss an employee's level and what is needed to advance.
- Payroll as a percentage of income is part of open book management with the team.

RESOURCES

American Animal Hospital Association (AAHA). Updated biannually. *Compensation and Benefits.* Lakewood: AAHA Press.

American Veterinary Medical Association (AVMA). Updated biannually. *AVMA Report on Veterinary Compensation.* Schaumburg: AVMA.

Veterinary Hospital Managers Association (VHMA). Updated biannually. *Compensation and Benefits Survey.* Alachua: VHMA.

Besides employee performance, what are other key factors to consider when giving an annual raise?

Long gone are the days when a person who simply survived another 12 months in a job automatically received a raise. Longevity is no longer a huge factor, partly because of payroll expenses, but also because we've realized that a team member who doesn't help to advance the practice and profit hasn't earned the raise. Just showing up and doing the same job this year as last year shouldn't necessarily result in extra compensation. We've also come to the conclusion that a person's performance should dictate how much money that employee takes home each pay period.

Performance can be measured and rewarded in several ways, such as moving up through a training program, learning a new skill or piece of equipment, getting certified, and successfully taking competency tests. Attendance and punctuality come into play here as well.

Factors other than performance also may be considered when deciding on a wage increase. When an employee is terminated, it is often not only that they cannot do the task for which they were hired, but that they cannot do the task while demonstrating the appropriate soft skills. For example, a technician who may not be able to put in an IV catheter initially, but can learn how to place the catheter, will be fine. That is someone who lacked the ability to do a task, but has the ability to learn. There are examples of these corresponding hard and soft skills all around the practice:

- Someone can bring a dog from the back to a client in the lobby, but also should smile, make eye contact, talk to the owner in a friendly way, and say "Thanks" and "Goodbye."
- Someone can schedule an appointment correctly in the computer appointment schedule, but should be able to admit having double-booked a time slot.

- Someone who needs another employee to help restrain an animal could ask, rather than order. The employee should smile, make eye contact, and say "Thank you" at the end of the procedure.
- Someone who is able to calm a cat enough to give her an injection is using only the appropriate amount of restraint.

If a team member cannot get along with patients, clients, and co-workers, it doesn't matter how great their technical skills are—you probably don't want to keep that person on your practice team. You want to reward team members who are creative, good at problem-solving, and have a positive attitude and willingness to help even when something goes beyond their own job descriptions.

Another big factor is someone's accomplishments over the last 12 months. Have they brought any new ideas or protocols into the practice that generated additional revenue? Have they increased client compliance, thus helping more pets and generating profit? Have they attended CE and then taught the rest of the team? Have they taken on a new task or responsibility, such as finding a way to enhance the office workflow, helping another employee with a task or providing new-hire training, or giving their time to the community in a way that reflects positively on the practice?

All of these subjective actions can be weighed when deciding whether an employee receives a raise. However, be sure to put such things into writing to avoid any appearance of favoritism. Just like other aspects of hiring, the basis for raises must be clearly described, as objective as possible, and fairly and consistently applied.

How can we recognize staff members without financial compensation?

When surveys of job satisfaction have been performed on veterinary personnel, the top answers are always a surprise ... because they do *not* include compensation. Many other factors rank as equally or more important to these professionals, such as respect, involvement, feeling appreciated, opportunities to learn and grow, benefits, and rewards and recognition.

That being the case, why do we automatically think it will cost money when it comes to recognizing team members for a job well done? These non-dollar factors cost nothing, except for the time and attention to provide them on a regular basis. Here are some good ideas to put into place:

- *Positive reinforcement:* Remind your team members that their positions are valuable. Each position has a purpose, and each person's contribution to the practice is appreciated.
- *Trust the team:* Resisting micromanaging and demonstrating a trust in your team members will enhance job satisfaction. It's difficult for employees to believe you mean "Good job" when you hover over them all the way through a project, rather than letting them take it and make it their own.
- *Continuing education (CE):* One of the reasons that veterinary medicine is so interesting and attractive is that it is constantly changing. There is always something to learn, and every team member has interests and passions that can be supported by CE. CE does not have to cost money to be a part of the practice; you can offer programs right there at the office.
- *Reaching goals:* When you learn about a team member's goal and help them achieve it, it can be energizing. Helping someone become a credentialed technician or veterinary

technician specialist demonstrates confidence in that person and makes the employee feel valued. Ask employees about their personal goals. Even if they are in school for something other than veterinary medicine, ask about how the class is going; remember if they had a big exam and ask them how it went. Show that you care about that person even when they are outside the walls of the practice.

It's believed that managers spend 80 percent of their time on their below-average employees—trying to coach them, evaluate them, train them—but only 20 percent on their topnotch people. Reverse this equation! Spend time on those who have earned the right to be on the team and gain your praise, and spend less time trying to rehabilitate a below-average employee.

Try to find something to thank each person for every day, but make it count. People will not question your intentions or sincerity when you take the time to thank them for something specific, such as rearranging the inventory cabinet or finishing a client education piece.

27

How do we get staff to understand that their pay is tied to the bottom line, but still keep them focused on quality care?

These two concepts are not mutually exclusive; the team can help to increase the bottom line *by* providing high-quality patient care. Quality patient care includes compliance, diagnostics, best treatment, rechecks, etc. All of these help to determine the pay of each team member. Yet, first, there must be some training of the team to show them where their paychecks come from, and where the money goes.

This training starts with fee setting and pricing. One reason team members may feel they are being short-changed is that they see invoices for substantial fees and costly products walk out the door, and they wonder why their wages aren't increasing in synch. In fact, they may not be in favor of the prices you charge. This can result in treatment plans being declined—the team member fails to get client compliance because a recommendation was made despite personal reservations, attitude, or outright omission of certain services or products. In any case, the client is not receiving information (either factual information or via a positive communication) to make an informed decision and comply with the final recommendation. In essence, if a team member doesn't agree that there is an equal exchange of valuable services and products for the money spent by the client, it will show.

Demonstrate what goes into setting a fee, for both products and services. Talk about ordering and holding/inventory costs, payroll for the time that each person spends discussing a product or service, overhead expenses (from office rent to equipment purchases or leases to advertising and taxes), and—oh, yes—profit. Then the discussion can turn toward the budget, and where the money needs to go once the client has paid. Using the activity below, help employees gain an appreciation for the cost of running a veterinary practice. Use a pie chart to show expense or cost categories and the respective amounts for each. It is important to let the team know that, if they can each help build a bigger pie, their own personal slices get bigger, too!

Talking about money once with the team is not enough; the conversation may have to continue. Using graphs and figures, keep the team informed of the money coming in and going out. Such "open book management" will provide insight into the decisions that must be made by the practice.

Finally, set goals for the team, such as tracking the number of boxes of heartworm preventive sold or services provided. Let employees know that providing services and selling products or medications are important not just to increasing the bottom line, but also to providing good patient care. That should create a sense of having a worthy crusade for the team to take part in.

Do It Now

Give each team member 100 pennies and a list of the main categories of the practice's budget (payroll, marketing, overhead, supplies, etc.). Let them arrange the 100 pennies in the way they think income is spread out among those categories. Have a team meeting or retreat to share results and compare with reality.

How do I offer benefits to my employees without breaking the bank?

What are "benefits" to you? Some benefits are paid time off, CE money, discounts on services for employee's pets, health insurance, dental and vision coverage, gym memberships—the list of benefits is endless, and that is a good thing, because it allows the business to get creative.

Certain benefits are mandated by law—minimum wage, Social Security, worker's compensation—and may depend on whether someone is full-time or part-time. Once you have determined the benefits that you *must* provide, you can look at getting creative about ones to add to attract and keep good employees. The previous question addressed recognition *without* dollar compensation. There are three things to consider in talking about benefits that put cash in the pockets of team members: successful operation of the business, appropriateness of the reward, and legal/accounting regulations. Even cash bonuses, though, don't have to break the bank.

Successful operation of the business means that the business can afford to do something special on occasion. You must charge enough for your services to cover the costs of operating the business. If you aren't bringing money in, you can't pay money out. Most veterinary practices set team wages as a percentage of income. Income goes up and the percentage looks pretty good; income drops and that percentage seems too high, so you begin to talk about cutting back, perhaps by rescheduling employees—but maybe you could find less-expensive rewards or benefits to offer as incentives.

Appropriateness of the reward ties together employee engagement and business success. You do not want to offer a reward that only benefits one of the stakeholders—a reward that pays out to the employees, but does not result in business income—or success will not be sustainable. This could mean a reward that benefits the employee but is a nightmare to implement (expensive to monitor, calculate, evaluate), which will eat up any financial benefit, or one that benefits only the business and creates so many hoops for the employee to jump through that it discourages, rather than encourages, the team.

The legal and accounting portion of benefits involves how much they are worth, what gets taxed, how it is reported, etc. Be sure to have a discussion with your accountant before starting a rewards plan, to make certain you are looking at the right numbers and reporting any payments properly.

The key take-home message about benefits is that, if you ensure they don't undercut business operations and they serve all parties, they won't break the bank.

Remember to educate the team about the rewards program, keep them updated on the numbers, and celebrate any successes. If you offer cash rewards, be certain to know what you want to get out of it and to reward for something specific and measurable. One of the problems of benefits/rewards is that most practices simply throw out money in the hopes that it will "stick" and have a positive effect on performance. Rather, a business needs to understand what it is trying to accomplish by rewarding the team for a behavior or result. It is easy to toss gift cards to the team at a monthly meeting—but the reward can soon become an entitlement, without any meaning or beneficial action. On the other hand, arranging for something specific and measurable means that the team has a goal, knows when they hit the goal, and reaps the rewards of success with a "gift"—in this case, a monetary gift.

How do I get employees to appreciate the fringe benefits they have and take these into consideration when examining hourly pay?

Communicating value is key—you have to let employees know the value of fringe benefits.

A yearly report on these fringe benefits is a great starting point, but don't just throw together a list of what you currently offer and present it to the team. Instead, take the time to make a strategic plan with goals to reach and build that into your benefits package for the year.

Giving everyone a summary of the benefits they receive opens a conversation about how those benefits are calculated, distributed, and if other ideas should be considered (see Question 40). Start a discussion about fringe benefits for this year—ask employees what they would like to receive, encourage new ideas, talk about ways to improve the performance of the business—and tie it all together by calculating potential fringe benefits that can be affected by team member involvement in business performance. The ideas can be everything from health-care contributions to incentive plans. Now the team isn't just looking at numbers as "So what—I got this 'stuff,' but it didn't pay my bills." Instead, they are actively involved in setting up a plan where they receive benefits that they value, and see the part they play and how business metrics relate to their benefits. Getting the team involved removes some problems associated with benefits/rewards that may have little meaning to certain members.

You need to communicate the connection and the value they receive—whether it is in additional take-home pay or additional benefits due to programs offered by the business.

Anytime an employee comes to you and says Clinic B pays better than you do, it is time to dig a little deeper into the real reason behind this situation. Are they aware of all their benefits and whether these are the same at Clinic B? Is the employee feeling undervalued? Are you paying them for their performance? Do they contribute to the practice and to the team? Oftentimes, employees who toss out the "pay up or else" threat are not really team members you want on your bus. The money is never enough for them and something else is always driving the dissatisfaction.

30

What is the fairest way to discount veterinary services for employees' pets? Family members?

Keeping things fair and equitable to the team and to the business should be the goal of any employee or family discounts. These must be legal, too—yes, the IRS has requirements for setting and reporting discounts. (Talk to your accountant regarding fringe benefits, especially the 20-percent discount cap, to exclude it from being reported as employee wages.) The options vary among all the different types of practices—there is no single answer to fit every practice and appease every employee.

A discount on services is a good benefit to offer, but you do need to communicate your program to the team and make certain they understand it, know how to charge each other out under the plan, and are aware of its limits. You want to avoid having one doctor do something for nothing for an employee's animal while another doctor follows the discount program, or an employee taking unfair advantage of the program. It is essential to put the discount program into writing, communicate it to everyone in the practice, make it part of your employee handbook, and involve the team in discussions about it.

Friends, family, neighbors … the list can go on. Some practices give a yearly amount to every team member to spend on pets as they choose. This amount is considered income and must be reported. Others state a specific discount percentage. Some purchase pet insurance for the team and make arrangements regarding payment of what is not covered by the insurance. Your goal is to develop a plan that is fair to all your employees and to the business, and is applied consistently, while providing excellent medical care fairly, ethically, and legally.

Giving discounted care may result in an employee taking in more animals than they can take care of. If an employee is adopting numerous pets and requesting medical care for them, it may place a financial burden on that employee and the practice, as well as create resentment if it looks like that employee is getting more out of the benefit than everyone else. It may make sense to set a limit on the number of pets for which an employee can receive discounts or on the total dollar value of the benefit. This is a topic to address and record in the employee handbook.

31

Do bonuses work?

Whether bonuses work can depend on what they are supposed to do for your practice. Routine, regular bonuses for no specific reason become viewed as an entitlement. When bonuses become expected and taken for granted, the benefit of offering them is lost.

The largest hurdle is to tie the bonus to the financial performance of the practice, to the employees' performance, or to a team's performance, but doing so is important. Establish benchmarks for performance (individual and business). Conduct regular performance appraisals and hold the employee accountable, being clear about what expectations are to be met by the employee before a bonus will be given.

Avoid the entitlement trap by addressing issues before you announce a bonus or incentive program. Begin with a clear goal that ties in with the business goals. Define start and end dates for the program. Determine what elements are going to be tracked and present updates to the team on a regular basis. Make certain that the program involves things the employee can control. Involve the team in sessions to create a list of incentives (different things motivate different people). Finally, communicate the program—put it in writing.

The Society for Human Resource Management (SHRM) has suggested developing something called the "spot bonus." This is a one-time, "on the spot" reward for an employee whose performance is outstanding, such as someone who has been working all shifts when the practice is short-handed or saves an animal that was not expected to survive. The amount is up to you and can be a flat dollar figure or percentage of annual salary. Spot bonuses are not a structured bonus program, but the business does have to set up a budget and a structure for the managers to use. The benefit to the business is that the spot bonus is not a guaranteed payout to anyone and everyone; it is based on a specific circumstance and a particular performance.

As with any bonus or incentive plan, you need to have a plan—a purpose for the spot bonus, what you hope to accomplish, the effect you want by offering the bonus, a specific reason to present it, etc. Simply handing out bonuses for the sake of making someone's day is really no bonus to the business at all.

32

How do I motivate practice veterinarians to embrace more management responsibilities without expecting more pay and relieve some of the management duties of the owner?

Only one or two people may legally own the practice, yet it's important for everyone to feel "ownership" of the business for it to thrive. This is especially true for associates (non-owner veterinarians), because team members will look to these veterinarians for leadership and appropriate behavior and performance. The "DVM" behind their names comes with a responsibility to project a positive role model for others to follow.

It also means that these associates may find themselves in the position of "managing" the practice, particularly when it comes to interactions between team members. This is due to their status on the team, not necessarily any formal, assigned managing duties. Overall, associates need to buy into the vision or mission of the practice, whether or not they are given management tasks. Everyone on the team, including associates, should recognize the big picture that, if the business thrives, each person thrives as well and gets the benefit of a successful organization.

When associates are given specific tasks of administration or management, they are being asked to go beyond what is required of them as veterinarians and leaders in the practice. Their involvement is important, and it also should be compensated fairly. According to *The Well-Managed Practice Report* from Witchett and Tumblin, practices spend an average of 3–5 percent of gross revenue for "management," which would include anyone in a management position and the time others spend on running the business.

Besides being leaders and role models, associates can help the practice in various ways. They could keep the website and social networking pages up to date and filled with current issues and discussions. Medical record auditing and templates can benefit from their input. Their opinions are important when it comes to setting prices, choosing and assessing employees and their performance, and setting schedules for the veterinarians and the team. While they may spend time doing "non-medical" tasks, and should be fairly compensated for doing so, it's important to note that the associates do not truly "manage" the practice; that is still in the hands of the owners and upper management.

As with any job-related matter, non-medical responsibilities of veterinarians should be included in or added to a job description, with the veterinarian's knowledge and acceptance.

RESOURCES

Blohowiak, D. W. 1992. *Mavericks! How to Lead Your Staff to Think Like Einstein, Create Like Da Vinci, and Invent Like Edison.* Homewood, IL: Business One Irwin.

IRS. 2012. "Publication 15-B (2012), Employer's Tax Guide to Fringe Benefits." http://www.irs.gov/publications/p15b/ar02.html.

CHAPTER 4: PERFORMANCE REVIEWS

Once you have hired the right people and trained them according to your expectations, it is time to determine how well they are performing. This is how team members are held accountable for their contributions to the practice. The evaluation of this contribution is the performance review, which has to be meaningful and not lend itself to micromanaging. It should reflect how people work both with and without supervision. How employees perform when management is not present is a good indicator of their initiative and self-control.

The performance review should contain questions that elicit responses demonstrating success in both the employee who is being evaluated and the peers who work alongside that person. Between formal reviews, there should be an ongoing conversation regarding the employee's goals.

How do you evaluate staff performance in a meaningful and measurable way?

The best way to evaluate team member performance is to have detailed job descriptions that can be used as checklists to compare expectations to actual behavior at work.

A solid performance management program can improve a practice's performance by improving employee performances. Having an educated and well-prepared manager to run the performance management program is the key to a successful program. Your manager must be educated on the type of program that will be used and the common problems associated with performance appraisals. Employees must be involved with their feedback. Having them conduct self-assessments is a great way to get their involvement, but should be a component of evaluation, not the only evaluation tool.

Your program should be unique to your practice—what works for one may not work for another.

There are various types of programs. Some are more like ranking systems; some use a mechanism called 360-degree feedback (obtaining feedback from the employee, his or her peers, and members of management); others are based on management by objectives and measure performance against specific goals; and others use a Behaviorally Anchored Rating Scale (BARS) that assesses behaviors. Whatever system you choose, there are some basic key elements to all of the types:

- Job descriptions, expectations, and standards that are communicated before starting the job and before the evaluation period
- Discussions between employee and manager regarding future career goals and training needs
- Discussions regarding strengths and weaknesses as related to performance, career development, training needs, and pay

- Performance criteria that accurately reflect the requirements of the position and include goals that are "Specific, Measurable, Attainable, Relevant, and Time-bound" (SMART)
- Feedback that is continuous, especially regarding weak performance.

Conducting the performance appraisal becomes even more difficult for the manager when problems become common at your practice. Managers need to be aware of issues such as lack of organizational commitment, lack of consistency, halo/horn effects, recency effects, favoritism, and failure to prepare. Any of these will affect the effectiveness of your performance management system.

Before conducting your next evaluation or appraisal, take time to strategically analyze your system and develop a performance management system that is understood and accepted by management so it is used in a consistent, fair, and legal manner. Then communicate the system to your team and begin to implement it. Remember that performance management isn't merely telling an employee all the things done wrong in the past year (corrections should be done at the time an error occurs) or a generic "good job"; it is about developing team members to achieve personal and business goals that will result in a win-win for all involved.

RESOURCES

Ashman, J.W. & Shell, S. 2001. *Play to Your Team's Strengths: The Manager's Guide to Boosting Innovation, Productivity, and Profitability.* Avon: F+W Media, Inc.

Performance Management Help Center by Bacal & Associates. http://performance-appraisals.org/faq/index.htm. Contains information about assessing performance.

Green, Marnie. October 2007. "10 Questions that Get Employees Talking During a Performance Evaluation." http://ezinearticles.com/?10-Questions-that-Get-Employees-Talking-During-a-Performance-Evaluation&id=799140

How do I create and maintain individual accountability without micromanaging?

How many of you have spent an afternoon with an employee, listening to the myriad reasons why they failed to perform their job as required? Some employees always see themselves as the victim—blaming others for their failure and looking to you to solve their problems. To get to accountability for personal actions and job performance, we have to take the matter back to the time before the employee was even hired.

It all goes back to business strategy—the way your business develops standards of performance, objectives, and consequences. This includes when hiring occurs, in hiring the right people, and by matching people to the right positions. It involves all those times when organizational goals were communicated—or not—to the team.

Again, the practice needs clearly stated job descriptions and an easily understood hierarchy of who reports to whom to start. Taking action in the following areas will improve both accountability and performance:

Performance management: Develop performance standards, align these standards to the goals of the organization, implement the standards (train, monitor, measure, and inform), and analyze performance against these standards. (See Question 33 for a discussion of evaluation types.)

Empowerment: Trust your team to get involved in decision making. Get their ideas and actively involve them in setting goals, tracking measurements, and assessing business performance (both the good and the poor). This is the open book management approach (see Question 27) mentioned previously. When employees have ideas for ways to do things differently or add new services, encourage them to write out proposals and then work with them to develop the projects. Aim to create an environment where it is psychologically safe for individuals to talk about ideas and mistakes, and give constructive criticism.

Consequences: Consequences do not have to always be negative—there are positive consequences for a job well done. Does your practice have both positive and negative consequences? We are all familiar with the negative consequences of poor performance—discipline, probation, termination, no raise, or no bonus. Do you follow through on these? Or do you threaten but never carry out? If you do the latter, then you do not have negative consequences. Good performance should be rewarded with positive consequences—bonuses, raises, new job responsibilities, or public acknowledgement.

Accountability occurs when the practice leaders set goals and directions, the team gets on board, standards and protocols are developed and communicated, trends are monitored, feedback is given, and consequences are used. In the end, you establish trust and see performance being maximized, and avoid the need for micromanaging.

How do you ensure performance by the team when there is no manager around, such as making sure they answer the phone as necessary, treat the animals appropriately, etc.?

When the cat's away, the mice may play! No matter how dedicated you are to the practice, there are times when you will not be present. This is especially true of a 24/7 practice, as shifts will be scheduled regardless of the presence of management; patients continue to need care, with or without you on the premises. Every manager also will have absences, such as vacations, sick days, family emergencies, and other situations where you will not be physically on the practice premises.

If you're worried about what your team is doing without you around, pop in or call in at random times to see how the phone is answered both in the lobby and in the back treatment area. Drop by the practice unexpectedly, just to see how things are going and to remind the team that you are always near. You don't want people to feel as if they're under surveillance, but you do want them to know that they are expected to perform at peak level regardless of whether you are onsite.

Even those tactics, however, cost you time and energy. If you are concerned that performance slacks off when there is no manager on the premises, it's time to consider middle management, or establishing a supervisor, lead, or head of an area. This person typically has good skills in a current position and ideally has shown management potential and expressed interest in a position such as this.

Introducing this person, and perhaps position, to the team has to be done carefully—you do *not* want team members to think you're setting up someone to "tattle" on them. The team must understand that this person knows their job, can be an advocate for what the position needs, will represent the group and its goals, and will provide a representative at management meetings and discussions.

Once someone has been appointed or promoted to supervisor, brainstorm on what has to be "managed" during the shift in question. What type of monitoring should be taking place? What type of situations should the supervisor or middle manager handle, and when should upper management be contacted? Do the team members on that shift understand what will be expected by the supervisor?

When the absence is temporary, such as vacation or leave, upper management will need to depend on delegating or "farming out" many of the duties the company needs to have done. For example, the office manager can take deposits to the bank, a supervisor can be shown how to process payroll, and the lead kennel person could handle the food order. In fact, having a back-up person for every vital role in the practice is not a bad idea.

Set up the right people to do the right job, and you will be able to relax when you're away from the practice!

What are the best questions to ask for effective performance reviews?

Open-ended questions—ones that can't be answered by a simple yes or no—are considered the ideal for eliciting good information, but there are probably as many performance review questions and forms as there are opinions. When you are developing the questions for your practice, do so with the thought of what you want to accomplish by asking the question, and how you might react to different answers.

A number of websites provide sample performance review questions. The information you want to get from the review will help you form the questions to ask. Having some "standard" questions helps with preparation and consistency, but be sure to develop questions pertinent to the current culture and future goals of the practice.

Suppose you ask, "Do you feel recognized and appreciated?" What will you do if the employee says "No"? You could instead ask how the person would like to be recognized/rewarded.

What do you hope to learn when you ask, "Do you think you're doing a good job here?" Instead, you could ask the employee to tell you what it means to them to do a great job or go above and beyond in their job.

Also be prepared for delivering hard news about poor performance. You may find you and your employee have very different opinions on the topic. Be open to hearing what employees think about their performance.

Another thing you can glean from performance reviews is information to use during strategic planning. If you are experiencing high turnover, you may want to ask questions regarding employee retention—"What will keep you here?" If you are experiencing a lack of growth in skills, you may want to ask what new skills employees want to develop (and how you can help them develop those skills).

You must also take into consideration whether the questions are being asked of the employee for a self-evaluation, or of a client, a co-worker, or a supervisor for a "360-review"—one that takes into account everything the employee is supposed to do and how results are seen by everyone with whom the employee interacts. Again, the Internet has a plethora of sites offering sample questions for this process.

It is a good idea to record the results of any performance discussions, and to put required actions and the timeframe for completing the actions into writing in the event of future legal review. Keep a copy of the performance review in the employee's personnel file.

37

Do self-evaluation tools work when used in conjunction with traditional reviews?

The self-evaluation is a crucial part of the "all-around" feedback compiled on each employee as part of a 360-degree review. Even if you use other methods of evaluation, a self-evaluation can become an important way to know the employee's concerns, future goals, and professional aspirations.

Too often, we fail to discuss the performance evaluation or appraisal process with a team member until they are due to receive such an evaluation. It is best, however, to explain the evaluation process at the beginning of the employee-employer relationship, either during the recruiting process or during the hiring process. Explain that in your practice self-evaluation is an important part of performance reviews, and ask the employee to reflect on the following questions as they go forward:

Have I completed my performance plan since the previous evaluation? Performance review should be a constant cycle of setting goals and achieving those goals. Ask employees to reflect back on their most recent set of goals, and assess how well they performed.

How will I improve on areas that were identified in the previous evaluation as needing to be done better? Each review time should also give the employee a glimpse at how to improve. It's important for the practice management to weigh in on what needs to be changed or done better before the evaluation process is over.

What were my achievements during this evaluation period? This is an opportunity for employees to brag! Encourage them to include CE they attended, information presented during in-house CE, additional training sessions they taught or took, personnel they trained during this period, ideas and/or suggestions they offered, and extra projects they may have initiated or completed during this period.

What do I want to learn or do next? Encourage team members to get creative and open their minds to the vast and changing world of veterinary medicine. Are there new protocols or procedures that interest them? Did they read about or see at CE another way to achieve success or monitor patients? The sky is the limit here!

What are my goals for the future—where do I want to be in the future, personally or professionally? It sometimes helps to put parameters on this question, such as "Where do you see yourself in 1 year, 3 years, 5 years?" This question can be pretty tough for team members, particularly when they are young and literally have their whole lives ahead of them. This question can involve goals related to the profession of veterinary medicine, but it can and should include personal goals as well. Team members want to feel that they are part of the family, especially because they spend a major part of their time in the workplace. Knowing an employee's personal goals can help a manager understand where that team member is in life, and where they want to be. This can help employees feel that the practice sees them as real people, not just cogs in the business wheel. Just be careful not to cross the line into questions that are illegal to ask.

In short, the self-evaluation should allow and encourage the team member to be thoughtful and reflect upon the past, be creative, and define the future, whether as part of a 360-degree review or a traditional review.

How do I offer an evaluation to a manager who does not take constructive criticism?

A manager who cannot accept constructive criticism should not be a manager. However, this question begs one to address two possible issues: the personality of the person receiving the criticism and the method of communicating the criticism. First and foremost, though, remember that the best way to make criticism easier to hear and accept is to start by praising what the employee is doing well before focusing on what has to be changed or improved.

Personality of the recipient: Regardless of the person's title (manager, technician, kennel worker, etc.), any member of the team should be open to constructive criticism. The ego can get in the way of accepting criticism and make a person defensive, even when people want to know if they are doing something wrong. Consider the age of the person as a factor to reckon with—it is said that millennials have a harder time taking criticism because they were raised during an era of praise. This does not mean that you should ignore this trait or forgive poor performance because of it, but it does mean that you have to realize the diversity of the workforce and that people of different age groups may respond to constructive criticism differently.

Some people do not realize how they respond to criticism. If that seems to be the case, aim for an open discussion to clear the air and help them see how to change their behavior.

Method of communication: The message must be clear and objective, addressing behaviors and explaining consequences to the team and the business. Are you telling the person how their behavior looks and how it is affecting the delivery of medical care or client service? Are you supporting your feedback with data?

Telling someone they have a poor attitude or are not managing people properly is not constructive criticism. Telling them the evening hours are too busy to let people arrive late for their shift and asking what steps they suggest taking to bring this to a halt is more specific to addressing the behavior and demonstrates that you are willing to help solve problems and map out specific changes in behavior that are needed.

In this type of situation, always look at all factors and address any gaps—either in an employee's acceptance of criticism or in your method of delivering the criticism.

The final answer may still be that the person is not suitable for a management position because, as we all know, the buck stops here—managers get the brunt of being held accountable. If they can't take the heat … well, you know how that saying goes.

How do we get middling performers to become high performers?

Many businesses have marginal performers. They would not be ranked a 9 or 10, but they wouldn't be called a 1 or 2. They are just—well—*there*. They do their jobs, but they never go above or beyond the parameters or show any initiative. There are a few points to take into consideration when assessing what to do with the marginal performer.

First, is this the correct person in the correct seat on the bus? That is, is the employee in the best job for the person? It might help to talk to the marginal employee about job satisfaction and what would result in better performance. A different position or change of duties might change that marginal person into a top performer.

The concept of strengths-based management can help here (Ashman, J. W. and Shelly, S., 2011)—by concentrating on a person's strengths, a manager coaches that person to continue to develop strengths, rather than trying to overcome weaknesses. Although fixing some weaknesses is necessary, it can also be a losing battle if you are trying to fit a round peg into a square hole. You wouldn't want to try to force someone with poor communications skills to answer the phone all day—you would give that person an assignment that doesn't require such interactions or, if the employee were willing, provide training in communications.

You also have to consider whether the bus is causing the problem. You may need to identify negative modeling in your organization. For instance, if you never reprimand or discipline tardy employees, then the culture of your bus is that it is OK to slack off and come to work late. This may transfer to other areas—it is OK to slack off on ... chart documentation, client service, etc. Check your bus as part of your initial assessment.

Second, is your performance management program up to speed? Does your program solicit and encompass feedback, goal setting, and training? All of these should be addressed in a performance management program. Be certain to clearly communicate performance objectives and expectations, identify any barriers, and set up SMART goals (see Question 33). Follow up on performance and behavior changes identified during the last feedback session with regular meetings that center around open discussions of the issues.

Third, are you providing ongoing feedback and acknowledgement? Motivation comes from within, but a good leader and manager knows that giving constructive feedback and acknowledging accomplishments go a long way in lighting the fire of motivation within a person. Additional motivation may come from assigning a coach or mentor (see Question 14), or getting the employee help with a non-work issue that is causing marginal performance.

If you adequately address all these points and still have a marginal employee, termination may be the only solution.

RESOURCE

Ashman, J.W. & Shell, S. 2001. *Play to Your Team's Strengths: The Manager's Guide to Boosting Innovation, Productivity, and Profitability.* Avon: F+W Media, Inc.

Should performance reviews be tied directly to wage reviews?

Historically, performance reviews and raises have been a pair, happening at the same time in most of our practices. Yet, we are beginning, as a profession, to realize that this may not be the *best* way to do it. Factors that affect the ability of the practice to raise wages may have little to do with the team members: an economy that is not so great, employee benefits that are not so cheap, needing to add to or reconstruct the team. An employee may have done a fine job and deserve a raise, but the practice revenue makes it impossible. If that's the case, level with your employees.

It's more difficult to get a raise nowadays. We are no longer rewarding employees simply for hanging around another year. Yet, not being able to give a raise should not stop you from generating and delivering performance reviews. People still want to know how they are performing. If someone is doing well and you can't afford a raise, look for other types of rewards—extra days off, an opportunity to be in charge of a special project, etc.

For performance/wage discussions to be separate, a foundation must be laid that gives the team a clear idea of the wage ranges for each position, and what must be done to take a step up the ladder. If someone remains in a position that only goes to $18 an hour at the top of the range, they shouldn't expect a $2-per-hour raise! Instead, give the person an outline of the training or experience required to move up to the next level, and therefore earn the raise.

The team needs to understand this concept from the first day of work. However, sometimes you have to backtrack and change policies or protocols that affect existing team members. If your practice has always done evaluations and given raises at the same time and you decide to change that process, sit everyone down and explain that this change is happening—including why and when.

If the practice is not able to give raises at all, regardless of performance, the team needs to have the short version of how the practice is doing regarding revenue and expenses. Silence is the precursor to gossip, dissension, fear, and a drop in morale.

While performance is a portion of the factors considered in an evaluation, it does not stand alone. One factor is the total compensation that the practice pays to employ a person, and that should be understood by the team. To help your team understand this, consider implementing a Total Compensation Worksheet (a form that identifies pay, but also the monetary value of all the fringe benefits and payroll costs for your practice). If you do so, explain the purpose and advantages of this new protocol to the team. Otherwise, they may feel that the practice is just making excuses for not giving raises.

A Total Compensation Worksheet should include the following:
- Payroll taxes
- Retirement plan
- Medical/dental/vision insurance
- Workers' compensation
- Disability/life insurance
- Bonuses
- Fringe benefits
  - Paid time off (vacation, sick leave, comp time, personal time)
  - Continuing education
  - Veterinary care discounts
  - Association fees/dues
  - Licensing fees

Ⅲ➡ **Do It Now**

Create a worksheet that lists all of the "hidden costs" associated with how much an employee costs the practice. Deliver this worksheet once or twice a year to each employee.

41

What's the best frequency for employee evaluations?

Typically, employee evaluations or performance reviews should occur 3 months into employment, then at least every year, or 12 months from date of hire, and that time period has become the norm in many practices. However, it might be smart to do performance reviews more often. The evaluation reflects the past, but really focuses on the future. If you wait a full year to do evaluations, poor behavior may take root and become established; if you do reviews at least every 6 months, it is easier to identify and fix performance issues before they become major.

The timing of reviews depends on how the employee's performance is being monitored and coached for those days between performance reviews. Once a year is fine—*if* you've been monitoring and managing that employee's performance all through that year; when mistakes happen, corrections have been issued, and retraining has been used at the time; if the employee's "attitude" has changed and that has been discussed as soon as it's been noticed; when interactions with co-workers have become hostile, the facts have been gathered from all sides, and the employee has been coached on better ways to handle the situation. In other words, issues and concerns (as well as rewards and recognition!) are not stored up over months and saved for the evaluation, but are addressed as they happen.

There is a saying that, when it's time for an evaluation, there should be no surprises. The employee should have a sense of what will be said and the owner or manager should be prepared to say it.

In a perfect world, this is exactly what you're doing—dealing with everything as it happens and keeping that door open with ongoing discussions. However, the world isn't perfect, and the manager or owner has more than just employee performance on the "to do" list. If you only sit down once a year with an employee to discuss performance and behavior, you are missing the boat. For one thing, it's difficult to recall everything that has happened with that employee over the year. For another, there will be surprises, and an evaluation meeting can take an ugly turn.

On the other hand, if you have an ongoing evaluation process (perhaps a 360-degree type of review, with input from the employee's self-evaluation and team members), it would be overwhelming to have evaluations more frequently.

In both of these cases, a "mini review" comes in handy. A shortened version of the larger process, a mini-review should occur every quarter or 3 months. It could involve quick input from a few peers, or just from the employee's immediate supervisor. The form for such a review itself is shorter and easier to fill out. The meeting may be brief, but at least you're connecting with that employee more frequently and building that relationship.

Certainly, if an employee has performance or behavior problems, and the issues are addressed right away, it's also important to schedule another, more-specific follow-up review at some interval (30 or 60 days). This interval is based on the type of infraction and how soon you would expect to notice change. It might also depend on time needed to retrain, coach, and counsel the employee.

What are some guidelines and ideas for meeting with employees in between performance reviews?

There may be limited opportunity for employees to sit and talk with their immediate supervisors between formal performance reviews. Days are busy with patients, evenings are busy with families and friends, so there has to be a focus on finding time for these conversations.

Regular, consistent meetings between the direct supervisor and the team member have to find a place on your schedule. Build them into the practice schedule so you—or whoever the supervisor might be—and the employee know they're coming and can plan accordingly; don't let them be surprises.

As these meetings do fall under the human resources umbrella, it's a good idea to "document, document, document" the meetings and what was discussed. Michael Gerber's "E-Myth" style of management discusses Employee Development Meetings (EDM) as time to help employees to be the best they can become. Using that concept, you can create a concise and useful EDM form that covers the following:

- *Purpose statement:* "The EDM is when relationships are strengthened, progress is discussed, and decisions are documented."
- *Agreed-upon goals from a previous meeting:*
  ○ Goals from the most recent full performance review or EDM should be reviewed during the meeting.
  ○ Goals that are agreed upon during this EDM meeting are added to the list.
- *For each goal discussed (new and/or existing goal), document:*
  ○ Progress toward or success in meeting the goal
  ○ Obstacle(s) to achieving goals or meeting the agreed-upon deadline

○ Required supervisor or manager assistance

○ Agreed-upon deadline for when the employee and manager feel the goal should be completed

Include a comment section in the form and a place for both the supervisor and the employee to sign and date the document. The form goes into the employee's personnel file until the next evaluation or EDM, which should be done at least quarterly.

One of the "new" goals during an EDM can be addressing an issue of poor performance or behavior. If the team member is having trouble getting to work on time, a goal can be set to have no "tardies" within the next 30 days. In essence, this form then becomes the first formal warning on the topic (a documented "verbal" warning).

The EDM can occur monthly, if supervisor and employee are able, but should be done quarterly at a minimum. The time period should not be tight; allow 30–45 minutes for each meeting. When you look at all the things you spend time doing on a regular basis, 4 hours (four quarterly, 1-hour meetings) is a small price to pay for the positive relationship you are creating and the potential for improving employee performance and morale.

Ⅲ➡ **Do It Now**

Create your own EDM form, and start using it.

The best team is one that is motivated, focused on the common good, and exhibiting good communication skills. It's important to encourage new hires, as well as the existing team members, to stay enthusiastic about their work and the practice. Some team members choose veterinary medicine as a career, and they should be supported. Other team members should be given the opportunity to make their "jobs"—what they do to pay the bills—into "careers"—what they do as a life's work—and see what the requirements are for this climb, but also respected if they are just passing through. Along the way, team members will undoubtedly be affected by personal issues, professional doubts, and even burnout and compassion fatigue. The practice can be the key to helping team members through these challenges.

What is the best first step in motivating employees to work as an enthusiastic team?

The process of developing an enthusiastic team has a few key steps; however, just implementing one step without the other may cause a breakdown in enthusiasm.

The best first step is to set clear expectations—let employees know what is expected of them, what the standards are, what the vision and mission are, what the goals are. Knowledgeable team members are more confident and more productive. Make sure everything is in a proper context—that team members understand the *why* of a procedure or strategy. Knowing why things are done helps employees feel invested in doing them right.

Get commitment—get the team involved; listen to them and value their input. Aim to build a culture of collaboration. Show them some numbers (open book management) (see Question 27) and develop objectives together (management by objectives). Having a culture where it is safe to share ideas and develop new projects pays off with continual involvement and improvement.

Then inspire and reward the team—rites, rituals, encouraging stories, rewards, and ceremonies all play a part in motivation and encouraging enthusiasm for the common goal. When a goal is reached, whether personal or business, celebrate! You don't have to hand out cash every time—even a simple recognition ceremony at a team meeting promotes the message that goals are important and the business appreciates everyone's accomplishments.

You are striving for a team that contributes to the success of the business. They do not have to be overly exuberant every day, but they do need to feel appreciated for their contributions, empowered to be responsible for their work, and committed to the mission of the practice.

How do we keep long-term staff members motivated?

Motivating the seasoned employee who has been on your team for a long time can be a challenge. Burnout is a risk that can occur at any time and to any person. Keeping someone motivated over the long haul requires attention to the individual, and figuring out what motivates each and every individual on your team can be a daunting task for any manager.

Help these team members develop professionally, offer continuing education and other rewards, get them involved in the business, and show that you trust them—trust may be the key ingredient in keeping someone motivated over the long term.

You probably have people who have been with you for years and show up day after day to perform the same job. To keep their energy and interest up, they may take on different tasks or projects; perhaps they have outside interests that connect with their current tasks. These clues can help you figure out what drives individual motivation.

Use your performance management program to your advantage in keeping long-term team members motivated. Take time to discuss their career goals and interests (see Question 45 for more on this topic). Paying attention to the individual—to someone's concerns, career development, and interests—is a huge factor in connecting with a person. A manager can make an even bigger connection by showing people they matter by identifying how each person, their jobs, and the practice group are all affiliated with a common goal. Belonging to a group that has similar goals, and having that group value one's role and input, will motivate that person to keep a commitment to and involvement with that group.

Some team members do not like to be in charge, but do like to play a role in business activities. Getting them to participate in a group project might be the way for them to feel motivated—they are not in charge, but their opinions and their involvement are important.

Offering to pay for continuing education is a good thing, but trusting that person to return to the business, present new ideas, and take charge of a project based on those ideas can really motivate your long-term team member.

Remember, it is all about *them*—and the key is for the manager to recognize this. It isn't all about the business; it is all about the key players on the team.

How can I encourage employees to consider this a career, rather than "just a job"?

Your team members have chosen a business—animal health. As a career, being part of that business can morph into other directions and areas of emphasis can change. Someone may join your practice as a tech and evolve into an educator at a tech school, then into a practice manager, and then into a consultant. A veterinarian who starts as a small animal practitioner could become a practice owner, and then morph into a pharmaceutical rep. Have any of these people abandoned their life's work in the career of animal health care? No— they merely changed their emphasis.

Consider your team. Are there ways that your practice can help some of them evolve—change their emphasis—simply by expanding the career box? Your practice can house their entire career lifespan if your practice has different areas for them to move into. Your practice can offer career opportunities to your team members and experience resulting win-win benefits:

- Empower the individual to grow and develop by learning a new skill, perhaps a skill that plays into a new service your practice would like to offer.
- Offer incentives to a team member to do community activities, such as science fairs, career lectures at schools, or animal health info sessions at local clubs.
- Map out a path for team members to get certified and then empower them to use that new knowledge and skill level in your practice to help the business grow.
- Provide networking to a mentor outside the practice who will enable the team member to grow personally as well as professionally.

A career is not about sitting still and becoming stagnant or complacent. It's about changing and growing. Some individuals tackle their career growth with gusto, attending conferences, taking exams, and creating projects and advancement paths for themselves. Others need a little help to keep them growing. Your practice can benefit by helping them grow professionally. Be an active advisor when someone is going to a conference and follow up on ideas they bring back that they feel would benefit them and the business. You can also plan strategically for career changes within your business and use the strengths of team members to identify areas of potential growth for your practice. Use your performance management program to identify potential career paths within your practice for team members who are committed to animal care as a career.

There will always be those employees, however, who are just coming to do a job and not creating a career for themselves in veterinary medicine. They may lack the initiative or the desire to make this a career and do not respond to encouragement from management. As long as they are performing well in their positions, your goal is simply to sustain that performance until they decide to leave or no longer fit in with the practice's vision.

What can be done to mitigate the trend of LVTs burning out after a few years, especially ER LVTs?

There is plenty of talk about burnout, and how prevalent it is in veterinary medicine. Burnout also has a close "cousin" that has to be considered—compassion fatigue. Either one can cause even the most dedicated employee to lose interest in actively contributing to or enjoying their work.

Burnout is considered to be the result of stress caused by a work environment. This environment can relate to any number of things: the location of the practice and how long someone's commute might be; whether the facility is new and polished, or run-down and shabby; and other factors such as pay, schedule, management, teammates, etc. Important factors in burnout, though, are overwork and repetition.

It is easy to see why a professional in veterinary medicine may feel burnout—the hours can be demanding, and the work can be physically and emotionally draining. Our particular work environment presents challenging factors—dogs that bite; cats that scratch; blood, guts, excretions; difficult clients; and more.

Even when someone is doing something they love in a great work environment, burnout can still happen. A practice can help employees reduce or head off burnout by being committed to the team member's future and longevity in the profession. The practice can provide support through having an adequate and well-trained team, providing recognition, rotating duties to vary work activities and emotional effects, delegating responsibility to give team members a focus besides the treatment work itself—even something as basic as ensuring regular hours and sufficient break or vacation time.

Compassion fatigue is a separate matter, but equally important to recognize. It is considered to be caused by the stress of the emotional connection we make with our clients and patients. While burnout is external, compassion fatigue lives deep inside our emotional experiences. It is a response to the emotional burden of caring—perhaps caring "too much"—for animals and their owners, about seeing animals in pain or dying. It is especially common among people in the field who have to put animals to sleep, or see tragedies and trauma as in emergency (ER) practices.

Leaving one practice but continuing to work in veterinary medicine will not provide the tools and opportunities to address compassion fatigue. Seeing no light at the end of the tunnel, someone with compassion fatigue is likely to leave the profession.

The practice can help the team member with compassion fatigue. Make sure each team member has time off to regroup and recharge their compassion. Discuss emotional events in the practice as a team, to explore any guilt, anger, or disappointment relevant to an incident. Don't let the same person handle all of the difficult cases. Consider bringing in professional help for colleagues who show the signs of compassion fatigue.

When veterinary professionals are blaming burnout for how they feel, it actually might be compassion fatigue. Awareness is key. Learn the signs of both, and be prepared to help team members cope when either becomes an issue.

Ⅲ➡ **Do It Now**

Do some research into the warning signs of burnout and compassion fatigue. Post the information in the break room, employee pay envelopes, or practice newsletter to help employees identify and resolve these feelings before they become overwhelming. Look into the possibility of establishing an employee wellness program or formal Employee Assistance Program (EAP) to offer counseling (see Question 50).

47

My staff seems judgmental about our clients. How can I change their critical language?

Team members complain about clients, managers complain about owners, techs complain about receptionists, clients complain about employees … it's part of human nature to complain. But when you hear complaining, especially about the clients who pay for the very existence of your business, it is time to dig a little deeper. It can mean that someone's needs aren't being met.

First step: Have a policy about gossiping and criticizing in public that is enforced with consistency. Link respect for clients to their value to the practice; remind employees that even the most difficult, disagreeable, or demanding client is still responsible for the revenue that pays employee salaries. If someone is making negative comments about a client, especially in a common area where other clients can hear, an immediate action by management is necessary, both to stop the activity right then and there, and to teach the employee the proper way to deal with their frustration.

Second step: Get the rest of the story. Is the employee experiencing frustration in trying to deal with a demanding client? Does a business rule make clients angry? Is there a legal issue, such as harassment or animal abuse? The potential "real reason" possibilities are many. This is where the manager gets ideas for the next team education session. Pin down the reason for the behavior so you can help team members cope professionally and comfortably.

Third step: Coaching, training, and brainstorming. Use those real-life scenarios to work with the team and help everyone learn how to deal with these situations. Some people are quick on their feet with a response; others need some tools in their communications toolbox.

Suppose a client says, "You're only in it for the money." Afterward, you hear employees in the front area talking about that "tight-wad Mrs. Jones" who drives the Hummer and complains about your prices. When you dig a little deeper, you may find any of the following as the real issue behind the team's complaining: They don't know how to respond to a difficult client; they agree and don't understand why a fee is so high; they don't like the fact that the client complains every time and ends up getting her way … Now you have something to work with—you aren't simply reprimanding your team and telling them to cease and desist, but can coach them on conflict resolution.

We often hear about "teachable moments." Overhearing employees complain about a difficult client is an opportunity to enhance customer relations and understanding of what makes your business run. Building a culture of trust with your team is of great benefit— the ideal is when your team can come to you and tell you how much they dislike Mrs. Jones (they usually ask if the business can "fire" Mrs. Jones), but that they want to work out this negative relationship. That puts your team in position to grow and the pet still be able to receive excellent medical care. Who knows—you may be able to turn around Mrs. Jones, too!

How do you motivate team members to understand business goals while keeping a fun atmosphere?

Understanding the business goals and needs doesn't have to be boring or demoralizing. Communicate business goals by providing the team with information (use open book management techniques) (see Question 27). Have regular meetings to discuss the goals but try to incorporate some fun into them through games, role-playing, contests, etc.

Another useful motivational technique is telling stories. Consider having a box for customer comments and compliments, and make reading those a part of every team meeting. Recounting stories of going above and beyond, or making someone's day, helps the team understand the standard they are to live up to, and acknowledges people who do good deeds. A good story is a power boost that can motivate the team—and one based on actual customer input underlines the business goals.

Taking time to celebrate is another great technique used by many people in the field. Consider celebrating personal goals that relate to the practice, such as achieving certification or learning a new skill, and personal ones, like a successful night on the local softball team. Such recognition builds self-esteem and contributes to a positive atmosphere.

Some practices take business goals and let teams get involved with creating innovative ideas to help the business achieve those goals. Team members learn something new, connect with others on a project team, and experience the reward of team acknowledgement when their goal is reached. Be sure to remember to recognize and reward—and demonstrate how the project benefits the business.

Look for fun activities to incorporate into the work day or week. People who are stressed, bored, or demoralized cannot be productive. A little bit of laughter releases endorphins and kicks up motivation for the day's work.

What are some cost-cutting ideas that do not hurt morale and culture?

Cutting costs without hurting morale and culture requires a two-pronged approach. One is discovering cost-saving measures; the other is applying morale-boosting techniques.

It's common knowledge that inventory and wages are the two highest expenses of any practice. We drive ourselves crazy negotiating the best prices with multiple pharmaceutical, laboratory, and other suppliers. We struggle with payroll as a percentage of income, trying to cut hours here and there to balance and manage it. We audit our charts and look for clues to where we are missing charges and how to ensure we are getting paid for everything we do. We reuse, recycle, and refurbish equipment—all in an effort to cut costs.

None of this is bad, but it can be demoralizing. People start to feel that the only thing that matters is reducing costs and saving money, rather than giving quality service. It helps to identify socio-psychological needs—three basic needs for a culture of motivation: learning, affiliation, and reaffirmation. To tackle cost-cutting without destroying morale, take some time to develop a culture of motivation. Most people have a basic need to expand their knowledge, skills, and abilities. Most people can be efficient and master a skill. Being part of a group gives us an identity and a positive experience of being affiliated with something important. Recognition of accomplishments provides affirmation.

It is possible to create such a culture and then use the momentum to tackle different issues in your practice. In this case, the issue is saving money. Educate your team via open book management (see Question 27) about the expenses the business faces on a regular basis (the need to learn) and encourage them to take ownership of various approaches or projects. Then use the natural tendency for team members to gather into groups to set up groups to tackle expenses. One group may find that recycling has some financial benefits; you now have a "green team" to reduce costs via recycling and reusing. Another group may have an idea to make appointments run more efficiently (capturing charges and taking less time for team members). Ask your inventory team to address efficient price-shopping and cut inventory expenses (answering the need for affiliation).

Educate your team, put them to work, and listen ... and then reward. Everyone likes to be recognized for their efforts. You can go from individual rewards to group rewards—from paying an individual team member a percentage of a cost reduction that person achieved to paying a quarterly bonus to the team for savings they identified.

Perhaps the biggest effort in addressing this question will be for leadership to encourage the team to suggest innovative ideas while assuring them that you value their ability to deliver quality medicine to every pet, every time, even while saving money or reducing expenses. Don't assume that team members know you appreciate their efforts; make it clear.

RESOURCE

Bacharach, Samuel B. 2006. *Keep Them on Your Side: Leading and Managing for Momentum.* Avon: Platinum Press, Adams Media/F&W Publications.

How do I get employees to focus on the job at hand and leave personal problems at home?

It is difficult when personal problems cause lapses in an employee's performance. We can do our best to keep employees headed in the right direction, with simple statements such as "I need you to ..." to help them stay focused. When the team understands the importance of putting these problems aside while at work, they can help redirect each other along the way, or let the administration know that a problem is out of their control. This situation may call for a meeting with the employee to review the goals of the practice and whom it serves, while tactfully asking if there is anything going on outside work that is affecting work performance.

It helps for the administration to lead by example, and refrain from bringing *their* personal problems to work.

And, the reality is, no one is two separate people—one at work, and a different one at home. We carry the combination of all our life's happenings with us wherever we go. The question could be, "How do we help employees remain dedicated to the practice and concentrate on their work while at work?"

First of all, we have to care what is happening to our employees, even at home (without being intrusive). If employees feel supported by the administration, that we understand and "get where employees are coming from," they will feel more appreciative of our support and respectful of our requests. In other words, an employee would probably work harder for a boss who says, "I know you're under a lot of pressure at home, but I really admire how you've been able to keep it together and perform well here at work" than one who says, "Hey, leave your problems at the door; you're here to do your job!"

The way to provide this support is to pay attention to the mood and behavior of your employees. If someone is clearly bringing personal problems to work (distracted, weepy, or angry; getting an unusual number of personal phone calls at the office; coming in late; performing noticeably less effectively), have a private meeting to ask how they're doing. If they tell you why their performance has been off, be understanding and ask how you can help—a change of shift, fewer or more hours, etc.—but be firm about performance being your priority. You are not supposed to be a counselor or psychiatrist for your employees. You aren't their parent, either. Your support, balanced by the need for everyone to pull their weight and do their job, is the most important message to get across.

It's also helpful to be able to provide your employees with help for personal problems. This is typically done by having an EAP. While we all understand the importance of health insurance, it is also important to recognize that your team members may need "mental health insurance" as well. An EAP is an easy way to determine whether someone needs counseling and how much they might need. Most EAPs provide a limited number of free sessions at no cost to the employee. However, the employer pays for this benefit for the team. Look at all your options from your team's health insurance provider, regional mental health offerings, and the AAHA-approved EAP website to compare prices and services.

RESOURCE

Smart Health Plans™. http://www.smarthealthplans.com. An AAHA-recommended EAP provider service.

51

How do you deal with friendships and cliques among employees?

While friendship can be the glue that holds the team together, allowing cliques within the workplace can pull that team apart. A clique can form even at a practice with only a few employees. This is typically a group of employees that picks and chooses whom they want to include in their circle. Those left out feel like outcasts, while the clique feels empowered.

Ironically, group activities outside of work can help the team get to know each other as people, not just employees, and help them bond and work together as a team. Yet, we have to prevent these relationships from coming between the employees and their work. When these relationships are strong and positive, the team can be successful; if there is conflict or hurt feelings, some team members will falter in their performance. While you have limited control over these relationships, you do have control over how much you tolerate poor performance. Just like any performance issue, the focus must remain on the tasks at hand and how well they are performed.

In fact, there's no reason to worry about these friendships if over-all performance is good. When the work becomes affected, though, this issue is no different from any other conflict in your practice. Meet with each person, or bring the people involved together and mediate a private conversation to help illuminate the issue. Let them know that cliques are not appropriate in the office and that you expect everyone to work together as one team. Give them time to improve their performance. If they do not meet the goals, they should receive disciplinary action just as with other performance problems. Focus on the what—what the employees are doing—without compromising the practice's vision.

How do we decrease turnover?

A small amount of turnover is healthy in an organization—it brings in new energy, ideas, and skills with every person hired—but excessive turnover can be a problem for many reasons. It costs the practice money in the time spent hiring and training people who leave, costs time in replacing them, can affect morale, and decreases performance of the entire team if there is a continuous revolving door at your practice.

When looking at turnover, it's important to recognize the difference between the positions in your practice. If you are hiring credentialed technicians, chances are they have already decided to stay in this profession. Your job is to make yours the practice where they want to stay.

Assistants or kennel help often are younger people who join the practice with both parties understanding that this may not be a lifelong career option. They may be working summers and weekends as they make their way through school. Some front desk team members are understood to be "stopping off" in veterinary medicine while they pursue another career, work their way through college, or take time out of an original profession to see "what they want to be when they grow up." Some people take a position that is open and offered, just to get their foot in the door for a job they want more—someone who takes a front-office or kennel job with the hopes of eventually being "in the back" with the patients. It's important for the administration to know this upfront, so they can support those employees by considering them when a preferred position opens up.

The best way to reduce excessive turnover is to have a formal exit interview with anyone who gives notice of quitting a job with your practice. Prepare a few questions to ask about why they are leaving, what might make them reconsider, what about a new position is more attractive, etc. If the only difference is salary, you may not be able to keep that person on board. If people are leaving because they don't feel valued or fulfilled, though, you need to know, so you can improve conditions for your remaining or new team members. It is also helpful to conduct "stay" interviews, to find out why team members are *staying* at your practice. This helps you to know what you're doing right in keeping good people on the team so you can keep doing it and maybe find ways to do it better.

To reduce the number of people who quit your practice, consider these factors in creating a positive practice culture:

- Consistently pay attention and offer feedback.
- Acknowledge a job done well.
- Keep practice financial status/goals transparent.
- Offer a comprehensive benefits package.
- Let team members use their technical abilities.
- Demonstrate respect and appreciate contributions to the team.
- Involve the team in decisions.
- Provide a comprehensive training program that instills confidence.
- Keep the management team positive and upbeat.

RESOURCE

Bacharach, S. B. 2006. *Keep Them On Your Side*. Avon: Platinum Press.

Where there is more than one person, there is the opportunity for conflict. It can be difficult to work with people who have different personalities and communication styles. Even when the team seems to be headed in a good direction, one disgruntled team member can throw the team into a tailspin. We expect a lot from our team members. We want them to come to us with both solutions and complaints. We want them to work out differences among themselves, yet to be aware of when it is time to involve management. If we give employees support and advice along the way, we can help them sort out problems that they can handle directly, and make better decisions about when it is best for owners or managers to be involved. From the beginning, team members need to understand that they are just as responsible for getting along with others as they are for their technical skills and knowledge. You might even want to draft a brief outline of how to respond to workplace conflict and make it part of the employee handbook.

What is the best way to handle conflicting personalities between staff members?

There is good conflict (identifying procedural snags or lack of knowledge) and there is bad conflict (diversity issues or power struggles). The manager's job is to observe, ask questions, and determine whether a conflict is a healthy one.

Begin by investigating. What are the core issues, what caused the conflict, what are the perspectives operating on both sides? Is this incivility (mere dislike or disagreement), or is it something bigger (discrimination or sexual harassment)? From the legal standpoint, it is usually a good idea to document what you find and what steps you take.

Talking it through is the best option. If a conflict is interfering with the smooth operation of the business, you have to step in. If team members refuse to get along, you may need to bring in a mediator to help both sides work through the conflict. Do not get caught in the trap of criticizing personality traits—these can't really be changed. Instead, identify the core issue and work toward the common goals of providing animal care, safe surgical flow, excellent client service, and the good of the business.

Conflict resolution involves articulating the problem and talking through ways to solve it peacefully. Mediation involves both sides agreeing on a solution, not the manager dictating a solution. The goal of both mediation and conflict resolution is to solve problems without anger, finger-pointing, or violence. This is where listening, brainstorming, and collaboration come into play. Consider asking both sides to write out their perceptions, issues, concerns, and suggested solutions, which can help the parties work through their own needs and frustrations before sitting down with each other.

Personality clashes are disruptive to the workplace. They may even escalate into aggression and violence. Management must take action when it becomes apparent that the conflict is affecting the workplace environment. It may be as simple as telling the two sides to cease and desist, or it may be serious enough to warn them that continued conflict will result in the termination of one or both parties. It is important to have a plan in place for managers to use in dealing with conflict and personality clashes. Develop your conflict resolution plan before you actually need it.

Do It Now

Research conflict resolution and mediation techniques. Your local bar association may be able to provide information and resources.

Develop a Conflict Resolution Plan and make it part of your manager's toolbox. This plan can contain forms to use to document investigations and conversations, recommended questions to ask, procedural steps to follow, action plans to get employees to help to solve their complaints, and follow-up actions.

54

How do we get the front and back of the hospital to work together more effectively?

Improving the way team members in different areas of your hospital interact is a matter of group mentality. The "we versus them" mentality originates from a very basic human need—to belong to a group, especially an active and powerful group. Owners or managers seek to hire team players and drum into employees that they are part of a team, so managers must take an active position to make belonging to the practice satisfy that need more than belonging to a small group or clique. This often starts with the first day of hire and the training the person receives. Leaving someone to wander around lost and confused opens the door for certain groups to take the newbie under their care; right or wrong, the new person now belongs to "the group" and hears the things the group believes.

The first step for a manager to take in dealing with a disjointed team is to explore the nature of the problem and the extent of group thinking. Next, change the work environment. Many practices cross-train team members in more than one skill or service, which can reduce the group mentality by enhancing team spirit. If cross-training is not feasible, job shadowing might achieve similar results. The goal is to get people to see the strengths of other team members and appreciate the value of their skills to the overall mission of the practice.

Management must also look inward when dealing with a group mentality, and consider the following:

- Is the work environment too stressful (extended hours, insufficient cover for shift shortages)?
- Do people feel threatened (money is tight, layoffs are imminent)?

- Has management caused the problem (recognizing one group more than another, not clarifying roles or emphasizing synergy)?
- Are team members not involved in decision-making processes of the practice (feeling left out of the mainstream, misunderstood, lacking the resources to perform their jobs)?
- Is there a conflict resolution process in place (not having an avenue to air grievances encourages like opinions to gather)?
- Are performance appraisals based on input from more than one person?

To improve the situation, managers can initiate team-building exercises that involve everyone in the practice. Instill pride in the mission of the practice and create a sense of being part of the greater good of the business objectives. An entrenched group mentality is difficult to break—it would be better to prevent that formation through activities that promote mutual respect, openness, and a common goal.

55

How do I stop employees from being constant complainers and help them solve their own problems?

Before you can stop employees from complaining all the time, you first have to understand *why* they are complaining. Some people just love to complain and need drama in their daily lives. Others complain in response to the stress of the work or a lack of clarity in job roles. Some may be responding to a misunderstanding in communications.

Complaints are most likely to center on working conditions, company policies, or personalities. When an employee is standing before you complaining, it is the manager's job to get to the root cause—legitimate or just a bad attitude—and take action.

Avoid playing the role of the parent who will "fix it and make it all better" for the employee. That only postpones dealing with the actual issue. The vital action for management to take is to listen carefully to the complaint and ask questions. Some schools of thought recommend asking "why" five times to get to the real issue.

Encourage employees to discuss their feelings and understand the real issue of their complaints. Explore the employee's complaint for any hidden agendas. Refocus the attention onto things they can influence and change (meaning themselves or their responses to the issues). Show that you have confidence in the complainer's ability to take steps to resolve the problem amicably.

If listening, understanding, and giving employees a chance to resolve problems themselves do not work, you may have to confront the chronic complainer. That negativity will destroy team morale and productivity. Make it clear that, if a complainer cannot stop the negativity, their job may be in jeopardy (see Question 58).

How do I deal with lack of respect between employees, especially with regard to positions with differing hierarchies?

What does respect look like? Consideration of others, consideration for diversity of beliefs and personalities, and good will to self. Respecting others means not making negative remarks that demean others, agreeing to help others, communicating openly, and appreciating the skills and knowledge of others. It also means give and take—you can disagree with someone, but you can express that disagreement politely and objectively, without name-calling, insults, or personal attacks. Lack of respect manifests as rumors, gossip, lack of cooperation, negative comments, passive-aggressive behavior, or even outright aggressive actions.

To reduce conflict over the hierarchy in the practice, provide employees with an organizational chart that makes it clear how everyone contributes to the business and emphasizes the value of every position.

We have all heard the theory that "to get respect, you have to give respect." When there is a lack of respect among the team members, the first place to look is at who is complaining and study how that person interacts with the team. Someone who criticizes colleagues in front of others, demeans others for their performance, or speaks poorly of someone's ability is probably not going to garner respect from others. Lack of respect also can show up in personal attacks for colleagues' personal or religious beliefs, personality traits, style of dress, and more—all inappropriate in an office team.

Respect for the hierarchy is always tough when promotions occur from within. It is difficult to be the buddy on the floor one month and then promoted to the boss of the team the next month.

Favors cannot be granted, and favoritism must not occur. The business should take proactive steps to train and guide a person who is going to be promoted from within the ranks to mitigate behavior or comments that could jeopardize the respect of those who were once buddies on the shift.

Most of our practices also have a wide age range of team members, and this comes with differing values that can lead to a lack of respect. Those born before 1945 are likely to consider self-sacrifice the norm for work. The millennials supposedly want to have fun in their workgroups. Gen X is said to want immediate praise and feedback, while the baby boomers consider giving a hard day's work enough fulfillment.

Improving how team members from different generations understand each other will lead to greater respect between them and a better appreciation of how the hierarchy works. This kind of understanding can be a factor in diversity training by making your team aware of these differences and how they can complement one another to make the practice successful and enable everyone to deliver the medical care pets need.

If there is lack of respect in your practice, individual situations must be evaluated, the entire culture of the practice must be examined, and corrective measures should be considered. Most generational misunderstandings can be resolved through better communication, but you may find that the one bad apple needs to be removed from the group.

57

How can we get our team members to stop blaming each other and focus on correcting problems?

It's hard to train team members in personal accountability for solving a problem in the middle of a situation. Employees who make mistakes often try to protect their egos or save face by resorting to finger-pointing and blaming others if they are confronted with having caused a problem. The best time to teach this skill is *before* the heat of a blame-game session.

The goal is a practice culture where it is psychologically safe to bring up differences of opinion or ideas, report errors and problems, and find solutions together. Start by having an open door policy to let the team know that they can bring concerns to you for open dialogue and problem-solving. Tell the team that you care more about the work getting done and people learning from mistakes than who made a mistake. Be sure to emphasize that blame is not the goal; learning and changing is the goal.

To build an empowered, self-directed, high-commitment team that can focus on correcting problems without casting blame requires teaching the skills for problem solving, communication, roles, and responsibilities. Educate the team on strategic thinking, how to present an issue without emotion, how to focus on possible solutions, and what role they can take in implementing solutions. Consistently remind staff, any time a problem occurs, that you are more interested in fixing the problem and preventing a recurrence than in blaming anyone. Involving the team in other decision-making situations in the practice will teach them the technique and encourage them to apply it. Focus on learning these skills and create a culture that can learn from mistakes and make improvements, rather than point fingers.

Managers must fight the impulse to "fix" everyone's problems and let them get away with blaming. As soon as you start emphasizing the steps for presenting, discussing, and solving a problem, the faster it will catch on, because others will see the positive effects of being in control of one's work environment. Once again, having this type of procedure in place and taught to all new hires will help when faced with an actual situation.

Perhaps most importantly, model non-blaming behavior yourself!

How do I encourage employees to try to problem-solve on their own?

Regardless of the practice, no one person should solve all problems. Sometimes this happens when the administration is micromanaging the team, or likes making all the rules. Any problem-solving the team members would want to do is crushed in this type of hierarchy.

We want people to be creative, offer solutions, and at times solve problems in the absence of a manager or supervisor. No one is continually watching every room in the practice, so every team member has to feel they can make their own decisions.

The owner or manager is a big part of this process. It's not hard to get bent out of shape when a team member makes a decision without us that we disagree with. How we react to situations like this will either encourage or diminish that team member's ability to apply solutions. You may have to step back, take a breath, and remember to be positive.

Ask for the team member's explanation of the event and the best way of addressing the problem. They should be able to tell you why their decision was favorable to either your patients, your clients, the team, or the business as a whole. Someone who is trying to live the mission of the practice should not be punished for taking initiative, even if a decision or action wasn't perfect. Acknowledge the solution and, if necessary, gently provide redirection, with language such as, "I understand why you solved it this way; I'm sure the client appreciated it. We want you to solve problems on your own; it's an important part of everyone's job. In the future, if I had the same situation, I would likely try … "

One way to encourage team members to step up and solve problems themselves is to include initiative and problem-solving in job descriptions, and reward employees who show those qualities with appreciation and, if it seems appropriate, some kind of bonus.

Make it clear that you will only listen to complaints or problems if the team member brings along a possible solution to discuss. This will empower them to take responsibility for the matter and ownership of the outcome.

There will be times when a team member's solution is not going to work, and you know it. Unless the solution is completely inappropriate or dangerous, let the team member move forward with it anyhow, so they benefit by learning from the mistake. Maintain a safe environment for people to approach with their solutions, and watch them learn!

59

How can I get employees to talk to each other to resolve minor conflicts and issues without involving me as the mediator on every single problem?

This problem is really about getting your team to take problems to their managers instead of you or other employees. A fine line separates these two issues. We want employees to feel that they can come to us—that our doors are open—but not for every little thing. We want them to talk to each other, but we don't want them gossiping. The right answer lies in the nature of the issue, the approach, and who is involved. It takes time to work these things out, and show the team the appropriate process.

As owner or manager, you do need to be involved in, or at least aware of, most of the issues that arise. When an issue is brought to you, tell the employee whether they should try to solve the problem on their own, or if you need to take it from there. Let them know that you have confidence they can resolve the issue, but if not, you would be available to mediate a meeting with both of the people involved, if it's that kind of matter. This avoids the "he said/she said" game, with the manager in the middle of trivial tattle-telling. It's also a good idea to create a "safe space" where the employees can talk together. A routine team meeting, without management present, can help the team develop its own solutions to problems.

It also helps to have a clearly defined hierarchy or organizational structure for the practice. Show in writing who reports to whom, and consider making a list of the types of issues that are likely to arise in the practice matched to who—or which job title—would handle them.

Tell employees that, if a colleague comes to them to discuss another person in the practice or a particular type of issue, they can redirect the colleague by saying, "You need to go to the people who can help you solve this," and walk away. This helps with gossip, too: If there is no one to listen, then gossip can't spread! Put this guideline in the employee manual to reinforce its importance.

⊪➡ **Do It Now**

Develop an organizational chart that shows who reports to whom by name or title and post it in the break room.

How do we stop established employees from picking on new employees?

There are phases of employment in every workplace. When a new employee comes onto the team, it can be a difficult time for everyone involved. Newcomers may feel as if they're back in school, being hazed or picked on until they find their place in the work "family." The new person is "green" and knows very little, if anything, about the practice culture and personalities of team members. This isn't the easiest time for the existing team, either. They need to make some room for the new person to fit in.

There is a lot for a new employee to learn, and that learning process will work better when a team welcomes the new person. You could assign long-time team members to be mentors or coaches for newcomers (but be sure to ask if the established employees want that responsibility; don't just assign it). Consider holding a welcome celebration, even if it's just cookies or donuts, for every new hire.

It helps if the existing employees can think back to their beginnings in the profession. Figley and Roop describe five phases of becoming a helper or caregiver that apply to this situation. The first phase is The Dream, when the person is so excited about the new position that it sustains them through their education or requirements. When they start their careers, they enter Phase Two, The Start—they are now helping the animals they love, and it is exhilarating. Existing employees may remember feeling that way themselves and may be a bit disappointed that they don't still feel that rush. They may be intolerant of the new person, and scorn such an upbeat attitude.

In Phase Three, Losing Our Breath, the new employee hits the wall of reality—the job may be more difficult than expected. The physical demands are difficult (lifting, restraining, standing, long hours, missed meals). They may question their career choice. In Phase Four, they are Desperately Seeking Rhythm—they will either figure out a way to survive, or leave the profession altogether. Finally, and ideally, the compassionate and increasingly skilled new employees will move on to Phase Five—Finding Our Rhythm.

Everyone benefits from employees who move through these phases and find their rhythm. To help the new employee reach that phase, create a welcoming environment. Provide not only a detailed job description and plenty of information about the practice, but opportunities for newbies to interact with and learn from established employees. Watch for signs of Phase Three and offer some company or advice if those appear. Remember how you felt way back when, and share this with the new person. Let newcomers express any negative emotions in a safe environment with someone who understands … you. Reassure them that every new job has a few bumps in the path and that the professional, physical, and personal elements will evolve and resolve.

RESOURCE

Figley, Charles R and Roop, Robert G. 2006. *Compassion Fatigue for the Animal-Care Community.* Washington, DC: Humane Society Press.

61

How do we teach veterinarians better conflict-resolution skills?

Veterinarians are in a difficult spot—they have to focus so much on animals and their treatments that they can become oblivious to the people around them. They may need help in developing better communication and conflict-resolution skills, especially if they're the ones creating conflict by overlooking the human needs of employees, colleagues, and clients.

It might be necessary to have a private meeting with a veterinarian whose communication style is causing conflicts; he or she may not even be aware of the problem. It can help to hold regular training sessions on communication and conflict resolution, using role-playing and tests like the Myers-Briggs personality analysis. Any of these are excellent options and will open the world of others to an individual to help them learn how to deal with people better on a daily basis. The trick is to make it applicable to daily work, and to keep repeating the knowledge gained and how to use it.

One effective option is awareness training or "habit training," along the lines of the behaviors of effective people propounded by Stephen Covey. Covey's Habits 4, 5, and 6 are especially relevant:

- *Habit 4:* Think win-wins, create clear goals and desired results. Explain consequences and state accountability.
- *Habit 5:* Seek to understand, separate the person from the behavior, and make sure you have heard and understood without judging.
- *Habit 6:* Synergize how the team is developed and treated by respecting and valuing strengths, participating in mutual problem-solving, and encouraging a creative team atmosphere.

Not only do these habits help improve co-worker relationships, they can easily be applied to client service, front- versus back-office conflict, and even personal or family relationships. We often feel frustration in dealing with people when we really got into the profession to deal with animals—but the animals usually come with a human attached, and the practice itself certainly does, so we need to develop "people" skills and knowledge as well as our animal skills, knowledge, and abilities.

Encourage both managers and veterinarians to attend conference sessions on the topic of communication and problem-solving to learn proactive, motivational approaches. Plan on a strategic planning session with conference attendees before the event—consider asking them to attend a specific session, perhaps based on the last performance appraisal.

Consider holding team meetings on the subjects of communication and conflict resolution; if you make them team-wide, the veterinarians won't feel targeted or put upon. Start planning now to give your team the opportunity to grow in their knowledge of people skills and develop the tools they need to meet situations head-on, rather than complaining about problems in the lunchroom or dumping them on the manager's desk. Proactive skill-building is the best answer.

RESOURCE

Covey, Stephen. 2004. *The Seven Habits of Highly Effective People.* New York: Simon & Schuster.

Do It Now

Develop a policy on conflict resolution for all team members and make it part of the employee manual.

How do I help employees understand that soft skills—personality and getting along with others—are as, or more, important than technical skills?

It's easy to communicate the required technical skills and workplace experience needed for an employee to be an effective team member, vital to your practice. Communicating the value of soft skills, like getting along with colleagues and clients, or just having a pleasant personality, is trickier. Team members may not know that these personality, communication, or attitude traits are important because the practice hasn't made its expectations clear.

These expectations should be spelled out in your fundamental HR tools, beginning with the job description and employee manual. The job description will describe the essential tasks, but also should include a list and definitions of the soft skills it will take to make the person successful in that position. While you can't control someone's basic personality, you can assess it in an interview setting and back up your impressions during an onsite visit, and not hire someone who demonstrates a prickly or difficult personality.

Here are a few qualities to look for in the appropriate category:
- *Knowledge:* language, body language
- *Skills:* active listening, speaking, active learning, instructing, critical thinking
- *Abilities:* written comprehension, sensitivity to problems
- *Work activities:* providing advice, assisting others, maintaining interpersonal relationships
- *Work styles:* compassion, integrity, friendly, cooperative, flexibility, accountability

Describe the type of person you want—someone who can mesh with the team, not just perform tasks. Use this job description to dictate the type of person you want—one who wants this type of culture in their workplace.

Often, what is important is not so much "how" a task is done in terms of mechanics, but how it should be done, including soft skills. For example, a veterinarian can know how to trim a dog's nails, but shouting at the assistant who is providing restraint while doing the work can cancel out how well those nails were trimmed. These soft skills need to be an integrated part of the training program, with trainers demonstrating the soft skills that the practice requires while colleagues perform the tasks. No matter how highly educated, veterinarians and managers should not be exempt from demonstrating respect to colleagues and clients through such soft skills.

When working to integrate these soft skills into the team, make it interactive; ask the team, "What does assisting others include? How about compassion—what does that look like? What do you see when you know someone is cooperative?" Finally, hold everyone accountable for these soft skills. On every performance evaluation, spend as much time or more on discussing the soft skills as on discussing technical and medical ones, and offer insights into how that person is measuring up. Peer reviews can be especially helpful, both because it's important that the team feels the team member has what is takes and because some employees take peer input better than management opinion.

You can use standard personality tests or assessments to make better hiring choices and management efforts in balancing various work interactions.

RESOURCES

Covey, Stephen R. 2004. *The Seven Habits of Highly Effective People.* New York: Simon & Schuster.

The Myers & Briggs Foundation. http://www.myersbriggs.org/. Contains information about using personality tests.

DISC Profile. http://www.discprofile.com/whatisdisc.htm. Contains information about Dominance, Influence, Steadiness, Conscientiousness (DISC) profiles and benefits of using them.

Once you have laid out the expectations of the practice and have given team members the opportunity to achieve success, there may be a need for discipline along the way. Through coaching and counseling, it may be possible to rehabilitate an employee who has veered off the path. How management handles the mistakes or shortcomings of team members is important to the level of confidence and safety they feel about divulging their mistakes or lack of knowledge. Sometimes, though, a relationship may have to end in termination. There is much to consider during these necessary but difficult aspects of managing people—legal and professional factors are involved, as are emotional factors for the member of management who must deliver the news, as well as the entire team.

How can we create a mistake-friendly environment?

You don't really want a "mistake-friendly" environment, but you *do* want one where team members are not afraid to admit to making a mistake, taking responsibility for the error, and learning from their actions.

People *are* going to make mistakes. Some will be minor; others will be serious. Is your business willing to accept mistakes and create a learning environment where these can be acknowledged and learning occurs? It is important not to let your practice become an organization that forces people to cover up their mistakes or blame others because the fear of retribution for the mistake is so terrible that team members cower in fear of being caught.

Robert Biswas-Diener has labeled mistakes as high-stake versus low-stake errors. High-stake errors are those involving safety or security (in other words, mistakes that put people or pets at risk or, worse, cause death). Low-stake errors are business mistakes that, although sometimes costly, point out areas for improvement in the business.

When considering what keeps team members from being accountable for their mistakes, a manager needs to look at the culture of the business. Many professionals say that being "free" to make low-stake mistakes—where employees are not on the chopping block when they make such mistakes—actually relaxes employees and results in innovation and higher productivity, because they are comfortable with learning and trying new ideas. Presenting these small mistakes at team meetings helps the whole team learn from them; just be careful not to seem to blame an individual—present the error, not the person who made the error. Creating such a culture involves having buy-in; establishing standard procedures; conducting training and monitoring; encouraging open communication; giving feedback and consequences—all culminating in the building of trust.

It helps to put the practice philosophy about mistakes into writing, perhaps in the employee manual. This could go near the section on practice vision, code of conduct, or teaching/learning environment. When new team members come on board, review this specific policy or section of the manual during orientation.

The other side of the coin is to look at the individual. People who are willing to take responsibility for their mistakes are also people who demand more from themselves and are committed to performing their jobs better. Some people may label this as a type of work ethic; it is a trait that may be underdeveloped in some individuals. Someone who continues to make mistakes and takes a laissez-faire attitude may never take responsibility and may need to be terminated. Of course, managers will want to proceed through the normal channels of meeting with the team member, developing SMART goals (see Question 33), monitoring and reviewing, and establishing regular follow-up meetings before releasing someone for their mistake.

Encouraging others to take responsibility for their mistakes and learn from them may be as simple as managers leading by example—admit your mistake, educate others about it, and move forward.

RESOURCES

Biswas-Diener, R. January 2012. "Embracing Errors." http://www.shrm.org/publications/hrmagazine/editorialcontent/2012/0112/pages/0112tools.aspx.

Society for Human Resource Management (SHRM). http://www.shrm.org. Contains a wide range of resources for hiring, managing, and firing.

How do I discipline an employee without legal consequences?

Call it what you want—constructive criticism, correction, reprimand, teachable moment—but any correction of a team member for making a mistake should be done in private and without anger or raised voices. Keep it professional. However, your action has to go one step further: It has to be constructive and legal. That means relevant to the job, based on a behavior, and in an open discussion with the employee about the effects and consequences of the behavior.

To protect your practice and yourself from legal consequences, it is essential to be proactive and have procedures and documents in place that outline what to do, how to do it, and the consequences of not following the rules of the practice. Be sure to apply the rules fairly and equally to all employees.

Never yell at employees, even if someone does something incredibly careless, and especially not in front of clients or other team members.

Start by gathering the information needed. Do not respond with a gut reaction; talk to all those involved and investigate the circumstances. Speak with the employee in private, discussing the circumstances; the effects on medical care, clients, co-workers, and business; the consequences for the behavior; and the employee's perspective on how to move forward. Outline the action plan—the steps needed to prevent a recurrence.

Some errors may call for termination based on the seriousness of the incident or the repetitive behavior of the employee. Having procedures in place (long before you have to reprimand or terminate) will provide managers with the necessary steps and documents for conducting either progressive discipline or termination due to gross misconduct.

As with any feedback or evaluation, document everything about any incident. Copies should be placed in the personnel file and given to the employee after a reprimand, because these documents may be needed as evidence in the event of litigation. There have been instances of lawsuits against a manager for perceived unfairness or violation of the Fair Labor Standards Act, Title VII, and other anti-discrimination acts. (A business can purchase Employer Practices Liability Insurance—EPLI—to cover employment law violations.) The key provision regarding reprimands and corrections: Proper procedures and documentation must be performed by everyone on the management team.

Do It Now

Put your policies and procedures into writing (or review them to see if they need updating) to create or update an employee manual, with a copy for everyone and at least one for the practice as a whole. Check with an HR attorney to make sure you haven't overlooked any important changes or requirements.

What is the best way to approach an employee who makes mistakes, knowing that positive reinforcement is important to building a good employee?

People who work in veterinary medicine are a caring bunch—we love animals and we love working with others who love animals. But we are also human and we make mistakes. Correcting errors and preventing recurrences is tied to the concept of teachable moments. Corrective actions should be done as soon as possible after a mistake occurs, and in private.

There are legal aspects to correcting an employee (see Question 64). Discussions should occur in private, where other employees do not hear or participate in the conversation, and there should be no defamatory or derogatory comments or name-calling. Information about the meeting cannot be disseminated to the rest of the team— that is, you can alert employees to the nature of a mistake and how to prevent it from recurring, but you cannot demonize the employee who made the mistake.

Be sure to maintain a level of concern for dignity—ask the employee what happened, what could be contributing to the problems being noticed, and how management could help, and give examples of the employee's strengths that could be used to overcome this particular weakness.

It is an excellent idea to get the person involved in fixing the problem. If you correct someone's mistake yourself, no one will learn from it and it is likely to recur. Sit down with the employee and discuss the error, troubleshoot the causes, outline preventive measures, and use the event as a teachable moment. Be aware that pulling an employee into the "fixing" procedure comes with risks. Managers need to have a partnership with the employee—both sides should show respect, dignity, and honest involvement in the process, and not take advantage of the process to vent and blame. Even a manager can get involved in blaming the "establishment" and feel as if their hands are tied as a way to avoid admitting defeat and not being able to help the employee solve the problem. Managers must know the role they are responsible for and not lose sight of the fact that team members need to take responsibility for their errors.

Don't let errors ride until an annual review meeting; address them immediately. Document all conversations about such events. Set SMART goals (see Question 33) to correct problems and prevent recurrences. Follow up and monitor how the employee is doing. Let practice members know that learning is a good thing and this is how correcting mistakes should be approached.

How do I balance treatment of employees when some are more productive than others?

People have all sorts of good traits and not-so-good traits, and even a great employee can have a bad day at work. Whether someone is your best or worst employee, it is essential to enforce workplace policies fairly and equally among all employees. Anything else will result in poor morale and could lead to legal problems as well.

The first way to ensure treating employees the same is to spell out what you expect in the employee handbook or policy manual. Be careful, though—do not include any policy that you will not enforce. That results in poor morale and inequality, and could lead to a wrongful termination suit. When you consider implementing a policy for the practice, a good litmus test for whether you will be able to enforce that policy is to hold it up against your "best" employee.

Let's say you have a technician who is always running late but is great at what he does: he can get a catheter in any vein, run a ventilator with his eyes closed, etc. You're not that sure that you want to risk losing this person because of the lateness issue. Ask yourself whether someone who doesn't care about following your policies is, in fact, a good employee. What will everyone think when this person gets away with being tardy—will they feel they can also come to work late? If another employee whose performance is only average is tardy just as often, would you terminate that person? If you do, how will everyone else feel about your level of fairness? Your policies must stand up to both the "best" and "worst" employee, to help you keep the people who are "good" all the way around.

It's also important to determine the impact of those best and worst employees, and how you manage them, on the practice as a whole. If negative things about this employee only affect a very narrow space—a small number of people, a non-critical task, or certain circumstances—and positive attributes far outweigh the negative, that is a different story from someone who has so many negative traits that the entire team is affected either through morale, work load, or the ability to be an effective team.

If you are worried about the effect on the team of terminating someone's employment, you have every reason to be—there are always ripples in the water when someone is let go (see Question 69). You may be worried that the people closest to that employee will be upset or angry, but they usually aren't. Most the time, they will have been aware that employee was not following the rules, but, because of their friendship, it wasn't discussed or addressed. You may be worried that the practice will be even more short-staffed, and everyone will topple from the extra workload, but that is hardly ever the case. The truth is, people who are more negative than positive affect every person on the team. With the negative person gone, there is often a refreshed outlook and ease of being with each other, and team members are usually willing to step up and work a little harder until you can fill the empty slot, because now the negative person isn't a sinker on the line—and everyone recognizes the consequences of poor behavior.

What is the right way to write up employees for disciplinary action or to document infractions?

A vital element of practice management is a process for disciplining employees who misbehave or fail to fulfill their responsibilities. Typically, a written warning follows a first verbal warning if someone does not meet the requirements of the job, except if immediate termination is needed for a serious grievance. Although it is called "verbal," this first warning should be documented and put in the employee's personnel file. You do not need an employee signature on a write-up of a verbal warning; you do need the person's signature on a formal written write-up. The verbal warning is an opportunity to show concern and ask if you can help, but also to make it clear that the behavior will not be tolerated. The written warning is the last step before you may have to move to serious repercussions.

In a typical "three strikes" discipline process, the first strike is the verbal warning, the second strike is the first written warning, and the third strike is the final warning. In the first and second strikes, you're working to help the employee rehabilitate, change behavior, and remain employed. The third strike is to inform the employee that the next time they behave this way, they will be terminated.

A template for written warnings can include some or all of these elements:

- Employee name and position
- Supervisor/manager name
- Date of incident
- Disciplinary level:
  - Verbal warning
  - Written warning
  - Investigatory leave (when theft or illegal activity is suspected)
  - Final written warning

- Subject of incident
- Category of incident:
  - Policy/procedure violation
  - Performance transgression
  - Behavior/conduct infraction
- Prior warnings, with dates and subjects
- Incident description:
  - Time, place, date of occurrence
  - Persons present
  - Impact on practice
- Performance improvement plan:
  - Measurable/tangible improvement goals
  - Training or special direction provided
  - Option for personal improvement plan from employee
- Outcomes and consequences:
  - Positive
  - Negative
  - Scheduled review date
  - Employee comments or rebuttal
- Employee acknowledgment:
  - At-will employment statement, if applicable
  - Acknowledgment of receipt
- Signatures and date:
  - Employee
  - Supervisor/manager
  - Witness, if employee refuses to sign

The benefits of such a performance correction template are many, but primarily it sets out a step-by-step process to use with all employees who may present disciplinary issues, helps managers deal with difficult situations objectively, and shows everyone that the practice treats them all fairly and equally.

It might be worth instituting a "decision-making leave" as part of your disciplinary process—at the final warning, you ask the employee to go home for the day, contemplate their employment, and show up the next day with either a letter of commitment outlining how they will make the necessary changes to stay on the team, or a letter of resignation.

The category of incident will help to prevent someone making it to a first warning for several different incidents. The inclusion of the "impact on practice" is important, but easy to forget. It reflects the reality that a poorly performing employee, especially one who goofs up but doesn't attempt to improve, creates resentment and problems with overall team morale, and can even cost the practice some of its clients/patients.

A template like this also gives you the opportunity to lay out methods and protocols for helping problem team members get back on track, and makes clear what will happen if an employee falls short again.

A witness is necessary only if the employee refuses to sign the required piece of paper, and they only need to witness the refusal; they do not need to know the subject or intent of the meeting.

RESOURCE

Falcone, Paul. 1999. *101 Sample Write-Ups for Documenting Employee Performance Problems.* New York: American Management Association.

How do I discharge an employee without repercussions to the practice?

When it becomes necessary to terminate an employee, two major topics must be considered: legal documentation and effects on the team. Any termination will affect the terminated employee, the rest of the team, and the business; a mishandled move can put the practice in legal jeopardy. Management's goal is to minimize the effects and mitigate any risks.

The Society for Human Resource Management (SHRM) regularly covers the subject of discharging an employee. The primary recommendation is to never summarily discharge someone (in other words, immediately discharge an employee without following official procedures), because that has both personal and legal consequences. To avoid this, before handing out a pink slip, conduct appropriate investigations or reviews; document which written company rules or policies were violated and actions taken in response; provide discipline, training, discussions, and reprimands; and produce further proper documentation.

There is always a potential legal minefield to contend with any time you discharge someone. Potential issues such as discrimination (see Question 82), constructive discharge, wrongful discharge, etc., must be evaluated during the process. This is where having a procedure already in place can help. By setting up guidelines for performance, reviews, investigations, and disciplinary procedures, a manager can monitor actions and steps through the entire process, from a first infraction to the termination, and protect the practice in the process.

The departure of a team member may either be a boon or a bomb to the team as a whole. It can be a blessing for someone to leave who is a poor performer or a negative influence. It is a shock if someone leaves who was popular, especially if the cause for discharge is not known—some people will feel the termination was unfair. In today's era of social media, many people will know about the termination as soon as it occurs, along with the negative feelings of the fired employee and any friends still in the practice. It is best to have a meeting with the team after a termination to give them a chance to voice fear, anger, or other concerns to the management team. This may occur as a team meeting, or an offer to meet one-on-one with those who have concerns. Just be careful not to say anything about the fired employee that could be construed as the basis for a later legal action—keep any explanation strictly limited to the facts.

Terminating a relationship is never easy. It can be made less frustrating by having procedures in place and a management team trained in the issues of legal termination and team morale.

If we terminate employees who are just getting by, are negative, or cause trouble, does it make the rest of the staff feel less loyal or insecure?

Firing someone has an effect on the whole team, even if the person being fired has been problematic. Team members may rejoice that a difficult colleague has left, but also may worry about whether they're next.

On any team, there are weak spots, such as those employees who are just doing enough to get by, and the team knows who they are. Typically, the management of the practice knows as well. These employees often also have negative attitudes and can cause trouble within the team. If that's the case, it's time to ask why you are keeping that person on the team, because their presence can affect the morale of the rest of the team. It could be that you're not ready or willing to enforce a policy, or you have not accumulated sufficient documentation to support the termination. You may hesitate if that employee is friends with others on the team. You don't know whether those people would support or condone your actions, but the risk is worth taking.

Generally, a team appreciates losing any "dead weight" that is dragging the entire team down. Even team members who are friends typically become annoyed or disheartened about a friend's problematic behavior. They will not want to look bad in front of the team because of this relationship.

If you hesitate to discharge someone because of the impact on everyone else's workload, recognize that this could become an issue—even briefly—but don't let it stop you from making a move that is good for the team. In reality, the team would rather see the person leave who is not carrying a fair load or helping out as much as possible. Typically, when the team loses that noncontributory employee, they will be energized and pick up the slack with no questions asked—there may not even be much of a load to take over! Initiate a search for a replacement team member as soon as you make the decision to discharge someone, so remaining team members don't feel they will be overburdened.

A side effect of this type of termination is the realization that each team member will be held to the standard expected by the practice leaders. Just as letting someone get away with bad behavior can lead others to follow suit, the opposite is also true—firing a poor performer will show the team that the practice will not tolerate mediocrity.

All that said, though, it is never a bad idea to reassure team members that their jobs are secure as long as they fulfill your expectations for performance.

70

How do you keep emotions out of the firing process, such as feeling bad about letting someone go because of their financial situation?

It is likely that you'll never forget the first time you have to terminate an employee; it can be a rather emotional event, because you realize that this will change the employee's life. If you are a compassionate employer, you can't help feeling for someone who is about to lose their income and their current livelihood. It's said that you don't fire people; they fire themselves by doing, or not doing, whatever is responsible for putting them in a position for you to fire them. But it still isn't quite that simple.

The most important thing you must do is look at yourself in the mirror, and ask: Did I communicate my expectations? Did I hold this person accountable for those expectations? Did we do our best to train this person? Did I provide coaching and counseling as needed? Have I provided ample opportunity for the employee to turn things around? Did I make these goals easy to measure and attainable?

If you can say yes to those questions, then you have no need to feel badly for that employee … you have done all you could.

This questioning process should help you cope with the emotions that may come along with the task of terminating an employee. If you've spelled out your expectations for and the repercussions of the employee's behavior, the termination should not be a surprise for either of you. You can assure yourself that the employee knew the results of a job loss and still did not choose to rehabilitate behavior or performance. It is human nature to be concerned, but you don't have to carry any guilt. Don't let your desire to make things right or be a caring employer make you deviate from the written declaration in the termination letter.

Often, the emotion that prevails is anger, either on your part or the employee's. If you are angry, you must wait until you can pull yourself together and not show that emotion to the departing employee or the rest of the team. You may not have any control over the employee's reaction, but you certainly have the responsibility to carry out this portion of your job in the most appropriate way. If the employee is angry about your decision, stick to the language used in the termination letter. Ask for a witness if you suspect this employee could be angry, or if the conversation starts to turn that way.

One way to reflect on the situation is to ask yourself, "If I had an opportunity to hire this person today, based on what I know now, would I?" Most importantly, remind yourself that you must take care of all members of the team and ensure a healthy business to continue helping pets and their people.

RESOURCE

Biswas-Diener, R. January 2012. "Embracing Errors." http://www.shrm.org/ publications/hrmagazine/editorialcontent/2012/0112/pages/0112tools.aspx.

CHAPTER 8: LEGAL QUESTIONS

Many aspects of owning and managing a veterinary practice involve requirements and considerations that are enforced by law at local, regional, state, and national levels. Neither good HR practices nor safety in the business happen by chance—there are rules to follow.

It is often challenging to stay current with all the regulatory agencies involved in our profession. Employment laws and regulations are constantly changing. Still, a manager must ensure that all employees follow mandatory safety procedures and company policies. Even the manner in which team members behave has to be monitored. From the incomplete personnel file to the expired employment law posters, the manager must be always vigilant in keeping the practice within certain parameters. Moreover, managers are responsible for ensuring the safety or privacy of their employees.

71

Is there an easy way to be aware of and stay on top of changes in employment laws?

It can be difficult to keep up with all the changes in employment law, yet the consequences of *not* being compliant are too severe to ignore. To complicate matters, the government changes the requirements— and lawsuits change precedents. Outside vendors that process payroll often have HR divisions that can answer your questions—check with your payroll company to see if you have, or can get, access to this service. If you want to manage your own HR functions, make sure your designated HR person is current with trends and issues in the field, perhaps by purchasing a Society for Human Resource Management (SHRM) membership for that individual. There also are attorneys who practice in HR and employee law, and are often invaluable partners in keeping your practice within the law.

One helpful step is to have all the relevant information in one place. SHRM has a wealth of information, including archives of laws listed by each state, as well as other tools and templates, etc. (Some information is only accessible to members.) AAHA and other veterinary organizations also provide guidance on HR and employment matters.

SHRM has chapters around the country, so check if your area has a local chapter of SHRM or another HR organization. Many hold monthly meetings to discuss changes in HR management, laws, etc. Non-members can usually attend local meetings to learn about issues facing local businesses, and to gain access to professionals who can provide information, connections, and consulting services.

Other websites offer information about legal requirements for businesses, but be aware that laws do vary by state.

Businesses in the health-care sector, including veterinary medicine, may be required to post certain types of information about employment, health, and safety. Various government agencies and websites provide lists of mandatory posters; you can easily access the list for your state by conducting a browser search. Some websites have links for receiving daily or weekly e-newsletters for updates on changes in laws or guidelines. Information presented by some of these e-newsletters can give your team a chance to discuss scenarios, get into the mode of thinking about laws and regulations (rather than acting on gut reactions), and create or update a manager's handbook with guidelines for handling various situations.

Be certain to have an employee handbook that has been examined by an HR attorney in your state for compliance with any laws or regulations. A professional can advise you on which laws and regulations you should or must include in your handbook and/or post in your practice.

It is difficult to know what we do not know, so we often start to question the laws when faced with a situation in our business. You may discover that your employees are more aware of HR-related laws and regulations than the practice managers!

RESOURCES

Society for Human Resource Management (SHRM). http://www.shrm.org. Contains a wide range of resources for hiring, managing, and firing.

Department of Labor. http://www.dol.gov/compliance/topics/posters.htm. List of government posters that must be accessible at a business.

HR Daily Advisor. http://www.blr.com. Newsletter with tips and updates on HR issues

Occupational Safety and Health Administration. http://www.osha.gov. Government guidelines on workplace safety and sources of workplace posters.

Ⅲ➡ Do It Now

Join SHRM. Consult one of the resources above to ensure your practice is in line with required information postings.

72

How do we stay compliant with the ever-increasing number of regulatory agencies that we fall under?

The most abstract, but a highly critical, topic in practice management is regulatory demands and employment law. The employer is required to know about laws at the federal and state levels affecting the business, and these can differ. One way to handle these complex matters is to divide up the task of staying up to date on regulatory matters and changes.

Typically, there is a department in state government for employment issues. The name of this agency can vary, but you should be able to identify the right agency through the U.S. Department of Labor. Contacting this agency can ensure that you are getting the state law information that you need. (Be wary of searching the Internet; a lot of businesses try to capitalize on helping businesses with the law, so be sure you use an official website.) See if your state agency has a newsletter or email update you could receive to stay up to date.

Belonging to organizations that need to stay up to date for their members is a good way to stay up to date yourself. Consider joining AAHA, AVMA, or your state's VMA. Their newsletters will contain news that you need to stay informed.

When it comes to employee safety, designating one employee as the safety officer can help your practice follow OSHA regulations. OSHA provides posters and publications aimed at helping businesses learn and comply with its requirements; these can be directed to the team member designated as safety officer and used to train and update the team on information they need.

Consider sending your designated safety officer to OSHA training every year, either onsite or online.

Various professional organizations often work with businesses at the state or local level to help them stay on top of both safety and employment law. You also can ask your practice or personal attorney, or local bar association, for help with getting the information and training you might need in that arena.

If you're in a large practice, the "divide and conquer" mentality can be taken even further. The office manager may be the best one to attend HR seminars and update the employment policy manual upon return; the safety officer would be the person to send to OSHA training.

Sending any of your team members to continuing education events not only ensures that you have current information, but also demonstrates your commitment to their role and the importance of the topic.

Do It Now

Join your state VMA, if you haven't already. Go to the OSHA website to identify current requirements and regulations.

73

How do we require, and carry out, drug testing and background checks on staff and new hires?

Any business in the medical field has to consider the reliability and safety of its employees, and that may mean administering drug tests and performing background checks. You need to know the legal guidelines on what, when, to whom, and what if.

You can tell a prospective employee that you will perform a background check before hiring, and that passing a drug test is a condition of employment. If you do either one without informing the candidate or team member beforehand, though, you may be in legal jeopardy. You also must tell *all* applicants about this policy.

Background investigations and reference checks are a means of confirming information provided by applicants. These may include criminal conviction checks, motor vehicle violations, poor credit history, or false information regarding work history or education. Establish your procedure with assistance from an attorney, because there are legal issues regarding background checks. To find and select a vendor who will conduct the check, ask local colleagues for recommendations.

Before initiating a drug testing policy, educate yourself on testing procedures and accuracy, as well as legal aspects.

Managers must be trained to recognize the signs of substance abuse, understand your policy on testing, and follow proper procedures for testing. Arrangements with a testing vendor must be made well in advance of needing to conduct testing. Although no legal requirements exist on what to test for, there are laws that regulate testing itself, so be sure to bring in legal advice when developing and instituting your policy. Make the policy part of the practice employment manual and, if it's a new addition to your policies, post it and give a copy to every employee.

There are various types of drug testing, but all must be explained to employees before being used. Pre-employment testing occurs after a contingent offer of employment is made. You should notify applicants that drug and alcohol testing will be required. Some states require that you give applicants a list of over-the-counter medications that may result in a positive test.

Once hired, an employee may be subject to other forms of drug testing as set up by the practice. "Reasonable suspicion" testing occurs when physical or behavioral symptoms are noticed, direct observation is made of use or possession, or a witness reports an incident. Random testing is just that—random. Employees may be chosen at random for testing—any employee, regardless of job title.

What if an employee tests positive? Specific actions must be followed. If you don't know what those actions are, consult your attorney before doing anything. Consequences must be in writing so employees know what to expect.

The goal of testing and checking is to create a safe workplace environment, where team members are qualified and able to perform to the best of their abilities. Managers and owners must be aware of legal issues and their own handbook statements when conducting these activities.

74

What does sexual harassment in the workplace cover— employees goofing off and hugging each other or making sexual references or performing inappropriate actions?

Sexual harassment in the workplace is illegal and is defined by the U.S. Equal Employment Opportunity Commission:

"It is unlawful to harass a person (an applicant or employee) because of that person's gender. Harassment can include unwelcome sexual advances, requests for sexual favors, and other verbal or physical harassment of a sexual nature."

Harassment does not have to be of a sexual nature, however, and can include offensive remarks about a person's sex. For example, it is illegal to harass a woman by making offensive comments about women in general.

Victim and harasser can be either a woman or a man, and the victim and harasser can be the same sex.

Although the law doesn't prohibit simple teasing, offhand comments, or isolated incidents that are not very serious, behavior is considered harassment—and illegal—when it is so frequent or severe that it creates a hostile or offensive work environment or when it results in an adverse employment decision (such as the victim being fired or demoted).

The harasser can be the victim's supervisor, a supervisor in another area, a co-worker, or someone who is not an employee of the employer, such as a client or customer.

This definition contains some gray areas, such as defining when an environment is considered hostile. This can be in the eye of the beholder, so to speak; it's possible that no one else feels harassment is taking place, but if an employee says it is, the organization must take immediate and serious action.

The topic of sexual harassment can also emerge in terms of office romances. HR professionals have differing opinions about whether this type of relationship should be officially put into a "love contract" to prevent future claims of harassment by either party, and to document expectations. These relationships can get particularly complicated when a supervisor is having a romance with a subordinate, which may be the most common connection.

If you do want to explore and possibly include a "love contract" policy—that is, guidelines on fraternizing or office romances—for your practice, seek advice from an employment or HR attorney with experience in these documents.

RESOURCE

U.S. Equal Employment Opportunity Commission. "Sexual Harassment." http://www.eeoc.gov/laws/types/sexual_harassment.cfm.

75

How do we get staff to turn in paperwork required for the job (e.g., reports required for the OSHA Log of Work-Related Injuries and Illnesses)?

To encourage team members to fill out and turn in required paperwork, make that activity a function of the job and make it clear that, without the paperwork, they are without a job. Sometimes we simply cannot allow an employee to dictate an outcome—meeting regulations that are required to keep the practice compliant and protect it against fines is certainly one of those times.

You can stress to the responsible party that filing that paperwork is not *just* a job requirement; it is essential to the practice being able to stay in business, so everyone's job—and the care of the animals everyone is devoted to—is at stake. That realization may help motivate someone to get on track, no matter how little they enjoy the task.

Provide examples of consequences to the practice that could result if paperwork is not complete: For instance, if an OSHA inspector came by, the practice would be fined for not having complete personnel records. Inspectors will not accept the excuse that the employee forgot or hasn't turned in the proper paperwork, nor should they be expected to.

You also could rotate the responsibility every few months, so no individual employee feels obligated to do this often drudgery-like task.

Steps that could make it easier to ensure the job gets done include:
- Present the paperwork and explain its function.
- Give the employee a firm deadline.
- Explain the consequences of failing to meet the deadline (suspension, dock in pay, demotion, etc.).
- Provide a reminder before that date, if possible. ˙

- Keep track of when employees turn in paperwork on time, and thank them.
- When employees do not meet a deadline, ask them why, *and* why they did not inform you of the delay.
- Give one last chance, but don't let it linger: If they fail again, initiate the consequences as outlined.

This may seem extreme, but a workforce often becomes lax about meeting certain deadlines and expectations because the administrators have not enforced the deadlines or demanded that expectations be met. It's human nature to procrastinate, but that does not mean you have to tolerate it on your team.

How do I comply with the Red Flag system?

The Red Flag Rules were created by the Federal Trade Commission to prevent identity theft. Although veterinarians were classified as one of the exempt professions, it is good business to address some aspects of the Red Flag Rules.

Establish a written policy with procedures that help your team prevent, detect, and mitigate identity theft. Your team should be trained on an annual basis regarding some of the relevant procedures to safeguard sensitive information and protect the business from liability. They should be trained on what constitutes a threat, how to respond to a "red flag," and how to handle confidential or sensitive information.

If you think it is silly to consider any employee of yours stealing a colleague's or client's identity and using it for fraud, just look around the office for opportunities to get someone's identity. There is the x-ray tag company report, listing employee names, birthdates, and Social Security numbers—sitting out for anyone to see. What about sweet Mrs. Jones, who wants you to keep her credit card information in her file and process her payments automatically—would a team member be inclined to use that card number for their own purchases? How many times has someone brought a sick pet in for treatment and said they were using their mother's Care Credit card? The situations are endless, and that is why having a written policy in place and conducting training sessions are important.

Even though veterinary hospitals are not required to follow the Red Flag Rules, it is still a good idea to have a policy in place, preventive measures established, and the team educated on identity issues. If a client's identity is breached, the client can be assured that you have measures in place to prevent a problem and deal with the issue.

If you do not have a written policy to deal with identity theft in your practice, do a little research or contact a veterinary consultant or identity theft consultant for information. Peace of mind is the name of this game.

Where can I download required postings without having to purchase them?

Certain information must be posted in every business, and every veterinary clinic or animal hospital. There are some options for obtaining required postings free, but they will require some of your time and you will need to consider the tradeoff—purchase the laminated versions or consider the expense of wages and printing in producing your own.

Check with your payroll company or insurance company for posters and Web links.

Each state lists its required posters and provides easy printouts. Unfortunately, there is no "one-size-fits-all" (just look at the differences between Pennsylvania and Ohio). However, the official state site also provides a concise list that is easy to print and use.

Many states have websites to connect you to your Department of Labor site.

The real question is whether you have discussed required postings with your team. We may place these posters in a conspicuous location, readily visible, where everyone passes by a dozen times a day—but does anyone take the time to actually look at them? Changes in requirements can give you a perfect opportunity for conducting team meetings to inform employees of their employment rights and obligations, and to carry the information further into discussions regarding procedures for conflict resolution, reporting discrimination, violations, or other infringements.

78

How do I ensure handling employee pregnancies correctly?

The Pregnancy Discrimination Act guides employers on actions they must take regarding pregnant employees. It boils down to, in general, doing nothing special. Employers should treat the pregnant employee the same as they would treat another employee with a temporary condition or disability.

The employer cannot force a pregnant employee to change her job or be terminated because of the pregnancy. The health of a pregnant employee must be considered, but, even though employers may feel they are looking out for a mother and fetus, forcing a transfer to another position or refusing to allow a pregnant employee to perform certain tasks, without the direct request of the pregnant employee, may be construed as discrimination or some other violation.

To take a proactive stance, an employer should do the following:

- Check your state and federal laws regarding pregnancy and taking leave.
- Include this topic in a section of your employee handbook (it could be under "Temporary Disability Plan").
- Have a procedure/protocol in place for your managers to follow when they are informed of a pregnancy or a temporary disability (for example, an employee who had a heart attack).
- Educate your team on the topics of employment rights and responsibilities, open-door policy, job performance requirements, and safety.
- Review your handbook to make sure it covers leaves of absence for such situations.
- Have the pregnant employee sign off on the review (especially when considering maternity leave, continuation of benefits, FMLA, etc.).

- Provide the employee with a written job description to take to her physician/OB-GYN for review and any recommended job modifications.
- Discuss job performance standards and ask that these be discussed with the employee's physician.
- Provide the employee with a written "hazards" letter detailing any risks of her position (x-ray, chemicals, lifting, toxic chemicals, or medications) and have her sign off on being aware of and taking full responsibility for any consequences.

If a pregnant employee is not performing up to standards or is in violation of workplace standards, treat the pregnant employee as you would any other employee who is not meeting performance expectations: Discuss, document, set up job improvement goals, monitor, and meet with the employee again (just like your regular protocol regarding performance standards). Early communication is the best policy. Your business will be in good shape by having policies and procedures already "on the books" before you are faced with an employee who has a temporary disability or is pregnant.

How do we afford the cost of workers' compensation insurance?

Workers' compensation (WC) is a cost of doing business that is mandated by the government. We should manage that cost like any other necessary business expense: by setting fees, generating income, and budgeting so we can afford to pay for it.

Workers' compensation is also affected by other regulations, such as the ADA and FMLA, so an injury on the job involves more than just that element. You can manage, and sometimes reduce, the cost by trying these techniques:

- Having strong safety guidelines and training
- Shopping around for a workers' compensation insurance carrier
- Understanding your incident ratings
- Taking proactive steps to educate your workforce
- Having policies in place to address the work status of an injured worker
- Working with your insurance carrier to mitigate the effects of claims on your premiums

Communicate with your WC insurance carrier to discuss your current status, the incident ratings, costs, and the meaning of some of the industry numbers. Look at how many bite injuries you have submitted to your WC carrier, or slip-and-falls, or back injuries from lifting, then work at changing your numbers. These numbers are a large part of your shopping-around experience and cost analysis.

Having a workplace safety program can be a major factor in controlling WC costs. Developing a safety committee can help your business "police" itself for risks. The committee can review injuries and conduct team training on preventing future injuries. Your insurance carrier and other organizations (such as AAHA, AVMA, and OSHA) have resources to help you set up a committee and/or educate the team.

Pennsylvania businesses can reduce their WC premiums by 5 percent by participating in the state's Workplace Safety Committee Certification Program. It requires taking specific steps, submitting required documents yearly, and holding regular committee meetings. You will need to determine if the committee produces benefits beyond the 5-percent savings. Check with your WC insurance carrier to see if special programs are available in your state or region.

Be sure to develop a return-to-work policy. Establish effective steps for handling an injury claim, modifying job duties, and reporting leave and documentation. Having good practices in place for documenting injuries and return-to-work will help the business deal with the expense of WC coverage.

What documents can and should be kept in employees' personnel files?

There are legal requirements for the documents that can or should be in a standard personnel file and—equally important—what cannot be in that file.

Specific items for a basic personnel file include the following:

- Job description
- Records relating to job offers, promotion, demotion, transfer, layoff, rates of pay and other forms of compensation, and education and training records
- Records relating to other employment practices (including policy acknowledgments and agreements)
- Letters of recognition
- Disciplinary notices or documents
- Performance reviews and goal-setting records
- Termination records

It's important to know who should have access to information on this employee. A direct supervisor is likely to need access to work-related documents, especially when evaluation periods come along. This would include documents that reflect the employee's performance, knowledge, skills, abilities, and/or behavior.

According to SHRM, documents that include protected and/or non-job-related information should be excluded, and filed where access is limited to people who have a need to know. These may include medical/health information, hiring records, and any investigation files if they exist.

These items *should not* be in a basic personnel file (but should be secured elsewhere):

- EEO/invitation to self-identify disability or veteran status records

- Interview notes and employment test results
- Reference/background checks
- Drug test results
- Immigration (I-9) forms
- Medical/insurance records (medical questionnaires, benefit enrollment forms and benefit claims, doctors' notes, accommodation requests, and leave-of-absence records)
- Child support/garnishment information
- Litigation documents
- Workers' compensation claims
- Investigation records

There is also the concept of exclusion when/if a government agency has reason to request your employee files. For example, if the Department of Immigration asks to look at your employees' I-9 forms, it is better to hand them a single file listing all employees, so they don't need to go through each personnel file. In this way, you're also stretching that concept of "need to know" beyond the practice.

81

Is it better to pay supervisors and managers on salary versus by hourly wages?

There are two models when it comes to paying someone to work for you: salary or hourly, and exempt or non-exempt. They can be combined in various ways, and it's very important to handle these correctly.

The terms "salary" and "hourly" describe the format or structure of the employee's pay. According to the Fair Labor Standards Act (FLSA), also known as the Wages and Hours Bill, a salary is a predetermined amount—a fixed amount that may not be reduced based on the quality or quantity of the work performed. Hourly is exactly that—the employee gets paid a certain amount for 40 hours per week, and then 1.5 times that rate per hour for any time worked beyond 40 hours per week. (Note: This is a federal guideline under the FLSA; check your state for any differences.)

"Exempt" and "non-exempt" are typically the trickiest terms. Exempt means that the employee is not entitled to the minimum wage or overtime pay. Non-exempt employees do receive overtime pay for time worked beyond 40 hours a week, or their defined term of employment, and in accordance with their state's requirements.

The FLSA has laid down the rules we must follow to determine whether someone's status is exempt or non-exempt. Executive, administrative, professional, and outside sales employees paid on a salary base are exempt from both the minimum wage and overtime provisions of the FLSA.

For FLSA Section 13(a)(1) exemptions to apply, an employee generally must be paid on a salary basis of no less than $455 per week and perform certain types of work that:

- is directly related to the management of his or her employer's business; or

- is directly related to the general business operations of his or her employer or the employer's clients; or
- requires specialized academic training for entry into a professional field; or
- is in the computer field; or
- is making sales away from his or her employer's place of business; or
- is in a recognized field of artistic or creative endeavor.

Care must be taken when applying the definitions above. The first two definitions would only encompass the main managers in the practice. The third definition—specialized academic training—does not apply to veterinary medicine at this time. The last three are not even close. How you want to pay or what would help the business the most is not what matters; these requirements are strictly enforced by the FLSA.

Managers can be on salary and exempt. Supervisors would need to be pretty high up in the hierarchy of the practice, and spend a considerable amount of time in management, to be salary and exempt. For everyone else, basically, you are required to pay overtime, which, in turn, means hours must be tracked.

RESOURCE

United States Department of Labor. "FLSA Overtime Security Advisor." http://www.dol.gov/elaws/esa/flsa/overtime/info.htm.

How do you dismiss an employee of a protected group without consequences?

When it comes to employment law, the most important thing to realize is that "protected" does not mean someone can never be fired—it means that you cannot terminate a person for being (among many other things) pregnant, over 40 years old, a certain race, a differing religious choice, etc. Even while protected under the Equal Employment Opportunity Commission (EEOC), people should be treated alike. The types of discrimination prohibited by the laws enforced by the EEOC include (and more can be added as laws change):

- Age
- Disability
- Equal pay/compensation
- Genetic information
- National origin
- Pregnancy
- Race/color
- Religion
- Retaliation
- Sex (i.e., gender)
- Sexual harassment

As long as the employer can show that the employee was not discriminated against for one of these reasons, the employer who follows the standard HR process should have no fear.

Keep in mind that termination is not all that is protected. Here is the complete statement:

"The law forbids discrimination when it comes to any aspect of employment, including hiring, firing, pay, job assignments,

*promotions, layoff, training, fringe benefits, and any other term
or condition of employment."*

It is important to realize that employment laws are not created
to protect business owners. They are created to protect the employee. If a case is brought to the EEOC that claims discrimination for
any of the reasons above, it is up to the employer to prove having
treated the employee just like any other employee. This is why "document, document, document" is such an important concept in HR.
All employees should be treated equally (which is exactly what the
EEOC protects), and the best—perhaps only—way to demonstrate
that equality is the documentation in the records of not just the
person who filed the complaint, but of every employee.

There is also a concept called "building a case," where an employer suddenly starts documenting problems with one employee and it
is not evident that the employee was a problem before a certain date.
If that employee is the only one whose file is suddenly getting lots of
documentation, it does not look good for the employer.

The most important element of protecting the business is to be
proactive. Create a foundation of HR documents, including a policy
manual, job descriptions, and performance reviews, and use these
fundamental documents to express the expectations and measure
the employee's success.

RESOURCE

Equal Employment Opportunity Commission. "Laws and Guidance." http://
www.eeoc.gov/laws/types/index.cfm.

CHAPTER 9: POLICIES AND PROCEDURES

Policies and procedures should be the easiest things for a business to monitor, because they are written down (or stored digitally) for the entire practice to review. Yet, some policies tend to be hot spots in human resource management—important, but often ignored by management. Is there a policy for the insidious gossip that can come between teammates? Is there a procedure for handling phone calls, and how employees are trained for that task? Even when good policies and procedures are in place, employees don't always follow them.

Policies and protocols protect the practice from the harmful effects of gossip, leaks in confidentiality, disagreements about early dismissals, and social media and online harassment, among other potential problems. They also establish expectations for employee behavior, from professional appearance and attire to giving discounts on services, keeping employees' own pets up to date with recommendations, and more—as has been said above, you must be aware of mandated policies and procedures and enforce them fairly and consistently.

Keep in mind that policies don't enforce themselves, and protocols don't do the training—they require manager enforcement and team member implementation.

How do I enforce a policy against gossiping in the workplace?

Gossip is almost unavoidable in any group of employees, but that doesn't mean it should be condoned or tolerated.

Most practices will develop some sort of gossip policy to help guide team members away from this potential problem. To be effective, the policy needs to define just what constitutes gossip in the practice. It could be any verbal behavior that is detrimental to the practice and its management, employees, or clients. This can cover a variety of behaviors. The policy must address what is to be done when a team member is found to be gossiping.

It is not always easy to find the gossiper. It is often times difficult to track back to who actually started the gossip, and who was involved along the way. You also have to define and circulate the repercussion—a warning? Termination?

It's a good idea to determine if there is a way to stop gossip before it starts. This means discovering why team members are gossiping in the first place. They may gossip because they don't feel safe going to management to discuss an incident or express a feeling, so be sure your open-door policy is alive and well, and supported by the leadership. Inequality can also fuel gossip. When it is perceived that one team member is treated differently from the others (whether better or worse), the whispers will start.

Another reason for gossip can simply be missing information. If there are gaps in the information being delivered to the team, gossip may start as a way to bridge that gap. If you communicate well with the team, they have no reason to look elsewhere for the information they need.

Another culprit could be fear, especially when the practice, region, or economy is faltering financially. If team members are worried about measures that might be taken in the practice, they start whispering about layoffs and terminations, which makes everyone uncomfortable. This is when open book management (see Question 27) and interactive discussions can help to ease the tension. You may have to confirm some of their fears, but it's better to know what to fear than not know what is going on. The team may even have some ideas for helping the practice cut expenses or increase revenue; if you make them part of the solution, they will be less likely to gossip about what is going on.

The receptionists in my practice have trouble avoiding discussions in the reception area, where clients can hear things that are not appropriate for them to hear. What are some strategies to deal with this?

To educate your team about the problem of letting clients overhear conversations they shouldn't, have them sit in different areas where your clients are going to be present. They should spend some time sitting in the front reception area listening to the chatter, watching the traffic flow, observing who is uncomfortable and why, and seeing how animals respond to the activity. They can also spend 15 minutes in the exam room and listen to all the conversations on the other side of the door. As a manager, you should try this activity periodically as well. You need to know if or where team members are likely to overhear sensitive medical information, complaints of clients or other team members, criticism of a client or doctor, etc.

This is a matter of privacy. To properly address this issue, you must have a privacy policy, and it must be provided and emphasized from the first day of hire and repeated on a regular basis in awareness training. Your customer service representatives or receptionist(s) in the front office are an integral part of the medical team and must be made aware that the medical and personal information of clients is a privacy issue and deserves their full attention and adherence to standard procedures. Ask the receptionist to brainstorm where and when there is an opportunity to chat privately.

Managers must respond to critical breaches of privacy. Call it a teachable moment, but when a receptionist is gossiping about that "crazy Mrs. Jones who just left here," someone needs to put a halt to it. A client complaint about what someone said in the back hallway should not be taken lightly, and the parties involved should be made aware of how their comments affect the client's perception of the reputation of the team and the business. People who repeatedly violate the privacy standard of your business may not be cut out for the job and may need to be reassigned or terminated.

How do we get our staff to buy into our core recommendations and follow them with their own pets?

Getting team members to follow practice recommendations for their own animals is all about mission and values. If the mission isn't known and supported or other values are communicated, the team will not buy in (and might even undermine practice recommendations to clients), so it is important for leaders to be aware of their important role.

Do you have team members who say they can't afford a dental cleaning for their animals even with their employee discount? What do you think they are conveying to your clients? This attitude needs to be investigated. If you simply told all employees that they can have as many free dental cleanings as they want, would they make more recommendations to clients? Or would you just be filling up your appointment book with free dental cleanings for employees? Explore other reasons your team members use for not keeping their own pets up to date, because these will be the same reasons you all hear from your clients: the cat stays inside, the dog gets too scared at the vet, etc. By addressing these issues with their own pets, your team members will learn how to discuss these obstacles with your clients.

Be careful about giving too many extrinsic rewards as a means of motivating the team to buy in to a task or idea. Extrinsic motivation is difficult to sustain and can lead to having to make the rewards bigger for motivation to occur. If there was no reward, would the team still perform the task or make the recommendation? If the answer is "no," you need to go back to the beginning to talk about purpose and involve the team in the development process.

Involving the team in some of the decisions of the practice via communication and open dialogue on the topic is a sure way to encourage engagement and buy-in. This does not mean letting anarchy rule, but instead, listening to suggestions and involving others in developing services and programs to monitor trends and educate the team. An easy way to coordinate all these efforts is to hold the team accountable. To accomplish this, the team has to know the "score" or what is at stake. A lack of buy-in can mean there is a lack of clarity in understanding what is at stake. Facilitating progress and recognizing contributions of team members goes a long way in getting the team to be fully engaged participants in the delivery of medical recommendations to their pets and to clients' pets.

Having a purpose is yet another aspect of getting buy-in. Make the team's work meaningful. Presenting a compelling vision with optimism and creditability helps team members grasp the purpose of an initiative. Don't ignore the power of leading by example. If the owner, manager, or other leader treats a recommendation as if it is not essential for their own pet, the rest of the team will continue that attitude with their pets and what they recommend to the clients.

Telling the team what's in it for them (WIIFT) is also necessary. Showing how a core recommendation benefits the pet, the client, the business, and the team members completes a circle and shows how good recommendations benefit everyone involved.

Buy-in requires motivation and momentum. You cannot expect to have buy-in if you are not communicating regularly with the team, holding people accountable for job performance, celebrating achievements, and leading by example, all of which contribute to motivating the team and keeping a positive momentum going.

What do you do when owners won't follow hospital policies, even though they have approved these policies?

A boss, owner, manager, or leader can't expect the team to behave a certain way if the very person responsible for setting the example does not behave in a consistent or ethical manner, and doesn't follow basic policies of the business. The owner sets the culture—others will eventually mirror the owner's behavior.

Owners often feel that, because they put their finances on the line and bear the brunt of the risks associated with owning a business, they can do whatever they want. They present a model of "Do as I say, not as I do." When the culture of the business takes a negative turn, the boss starts pointing fingers at everyone else. The culture will spin out of control and the business will suffer the consequences.

Bringing this to the attention of the boss is not without risk—to your reputation, your job, or your relationship. Most people do not want to hear what they should stop doing or start doing. They have to come to terms with themselves and want to make some changes because their situation has become painful. Personal behavior changes are difficult.

However, there is another way to explore this topic.

Is the boss breaking a rule because it is interfering with achieving a goal or other objective? Perhaps it is time to sit down and discuss the rule, to determine whether it has become obsolete. Maybe the rule needs to be changed. Perhaps medicine, laws, or circumstances have changed and the rule no longer makes sense. An example may be a medical protocol that was set up a few years ago. The associates are following the rule, but the boss is not. The boss may have read about a new technique and didn't think to communicate it to everyone else.

If the boss is open to change and to realizing that constantly breaking rules is damaging the practice culture, there is room to work on the situation and efforts can be made to address the rules and behavior. Team members may have to call a general meeting to brainstorm about the situation and get everyone's input on the nature of the problem and ways to resolve it. If, on the other hand, the boss is unlikely to change, it may be time to find a new boss whom you can respect.

RESOURCE

Lee, Fred. 2009. *If Disney Ran Your Hospital.* Bozeman: Second River Healthcare Press.

How are you most likely to succeed in implementing new protocols?

Hundreds of books have been written about change management, yet we continue to have problems getting change to occur in the workplace. There are definite steps to follow when implementing a new protocol in your practice:

- *Provide clear information:* Hold team meetings, discuss the goals for the new protocol, put it in writing, distribute it, and provide step-by-step instructions and training.

- *Demonstrate urgency:* Feel the need for change, show why you are implementing the new protocol and how it will benefit pets and the business, brainstorm and take feedback, address fears, and engage the team in being part of the change.

- *Monitor the process:* Show appreciation for what is working, welcome suggestions on how to expand what is working, make it easy—remove barriers, celebrate successes, and connect with successful "change agents" to mentor others.

The veterinary profession is part of a changing industry—new drugs, new procedures, new tests, and new diseases arise constantly. Your practice cannot sit still; it must be ready to change. Taking the time to develop change initiatives is the first hurdle; once you taste success, the next change initiative becomes a little easier to handle.

Be aware that mishandled change initiatives can result in poor morale and low productivity. Lack of executive support will kill any change initiatives, so be sure to get the involvement of team leaders. Analyze the need for change by knowing your gaps and weaknesses—don't change for the sake of change, but because change is needed and the new process will make things better in some way.

Many of these steps are about communication—from communicating at the beginning of the process to communicating with team members along the way to getting feedback about how the process is going. The real key to effective change is communicating the need or urgency for it and the problems that the team might experience by not changing.

How can I help employees understand that, if they can't find "busy work," someone has to go home early when we are having a slow day?

It's hard to get employees to buy into the value of "busy work" or the rationale for sending someone home when business is slow, especially if the latter means the employee loses income as a result. Before employees can understand, they have to have information that will lead to that understanding. All too often, managers and leaders know much more about the financial health of the practice than the support team does.

If the team has information about the basic expenses for the practice, such as marketing, inventory, lease or mortgage, operating costs, supplies and equipment, employee benefits, and payroll, they will be more likely to pitch in when savings must be made. The cost to keep the practice open for one day can be calculated and discussed with the team. Explain that, if employees are being paid when they have nothing productive to do, it can cost the company money and make the pie smaller for everyone.

To head off this problem, have a list of tasks that can be done at slow times. Let idle employees choose to either work on one of these tasks or go home early. Make sure these tasks lead to an improved practice, and are not just busy work. These should be tasks that you'll need someone to do in the next few months anyhow, whether or not business is slow. There are always tasks to be done that can directly improve the revenue of the practice—making reminder and follow-up calls to clients, running reports, calling patients who haven't been seen in 6 months, updating the website, working on newsletter articles or blog posts, updating files, cleaning, spending time on continuing education, checking inventory, etc. Give the team ownership of the issue by asking them for other ideas of tasks that can be done when the pace slows down.

⫸ **Do It Now**

Come up with a list of tasks that can be done to fill time when business is slow. Put a "slow time" policy into place.

How can we discourage absences and reduce excessive sick time?

To discourage frequent absences and reduce excessive use of sick leave, have a policy and follow that policy. It will not help the business or the team if those who ignore the policy are allowed to keep breaking the rules without consequences.

Do an Internet search to see what creative ideas are out there regarding calling off work. Here are some:

- Put accepted reasons for absence in the employee manual. Include requirements, such as a doctor's note for several days missed.

- Give bonuses to those with great attendance. It might seem like reverting to grade school, but adults appreciate it when their work ethic and reliability is acknowledged.

- Make team members responsible for covering their own shifts; peer pressure may come into play for someone who keeps calling in absences.

- Schedule an extra person every shift; if everyone shows up, send home the one who calls off the most.

- Reduce hours on the person's next schedule until they set a pattern of showing up for work as expected.

- Have a policy of allowing a set number of call-offs in a 30-day timeframe with consequences for violating it.

- Instead of having designated sick days and vacation days, provide for general paid time off that can be used anytime an employee calls off for the day. This has been shown to build trust because the team member does not have to lie in order to use a "sick" day.

It can be frustrating to deal with an employee with a habit of calling in to skip work. To add to this frustration, there are certain legal issues to be aware of: the Family Medical Leave Act and American Disability Act, for starters. Be sure to educate everyone in the practice about these protections, especially anyone with responsibility for hiring, management, discipline, or firing.

Managers should discuss frequent call-offs with employees and assess whether regulations apply to your business and the situation. Try to determine whether someone is making frequent call-offs because something is affecting their commitment to the practice. Sometimes, call-offs are a sign of a deeper problem with team morale, practice culture, compassion fatigue, burnout, or family situations. These should be examined on a case-by-case basis and appropriate referrals for assistance then made.

Many businesses have established Employee Assistance Programs (EAP) to assist employees and their immediate family members with healthcare issues. The goal of an EAP is to help the employee manage personal issues that may be affecting their work performance. Many of these programs (all conducted on a confidential basis) include counseling, assessments, evaluations, and support on a variety of issues (see Question 50).

Let employees know that excessive absences and call-offs hurt the business. They hurt morale, they hurt the delivery of medical care, and they hurt client service. The challenge for managers is to investigate the root causes, discipline those who need it, and assist the others who have a need.

How do you get associates to value their services and skills and reduce discounting?

Doctors giving discounts is a difficult issue to tackle. You would hope that associate veterinarians feel a sense of ownership in the success of the practice, but even owners sometimes are quick to discount. They are setting a bad example for the other veterinarians, so owner backing needs to be in place before you tighten up on discounting.

Before tackling this issue, do some investigation. Establish raw data on how much discounting has taken place over a certain time period for each doctor. This may involve a brief form to be used stating doctor name, client/patient name, amount, and why the discount was given. There are many reasons, so your tactics may change based on these reasons. If they don't value their services and skills, this is a self-esteem and self-confidence issue that may require private time to work on. If they think the practice is overcharging for services and products, educate them on what is involved in setting fees. Make it clear that it is *their* profit they are taking away from the practice, and themselves.

Some veterinarians may think that they could help a pet owner more easily afford your prices; you may wish to establish a Compassion Fee for those situations. The amount that has been deducted or removed should be apparent to the client on the invoice so the client can see the value of the adjustment. If this happens regularly, it might be a good idea to establish a charitable fund to help those who cannot afford pet care.

Some practices take the doctors out of the equation entirely when it comes to fees and discounts. The support team generates estimates, presents them to clients, and enters charges. The support team members who provide the actual hands-on labor are likely to keep those charges on there because they know they earned those fees!

It is imperative that the practice have a written policy on discounts—amounts, reasons, and personnel allowed to offer them—and that the information be distributed to those who are discounting. Create a policy, and then stick with it! Make it known to associates during the interviewing process, so they can make a choice regarding this facet of employment before joining—and possibly undermining the profits of—the practice.

How (and can) we control what is said by staff (current and previous) in social media about our facility?

It seems that, every time we turn around, there are new types of media out there, and we have to figure out how to protect the practice from what people can say about our organizations. It is imperative to have a social networking policy these days, because social media can be both a marketing tool to improve the practice and a forum for grievances to be aired online.

You can't prevent employees from taking part in social media—it would be futile anyway, because you can't monitor behavior outside work hours and away from the practice. However, the practice can strive to protect itself from unauthorized disclosure of information. You can establish a policy that employees are prohibited from speaking on behalf of the practice. This includes information about clients, patients, products, fees, employees, or any work-related matter. If a policy exists and someone violates it, you can terminate them for doing so. You may have limited protection against someone outside the practice making damaging comments, but legal action for slander and libel may be an option.

Employees should be told that they should have no expectation of privacy while using the company's equipment or facilities, and that the company has the right to monitor this on behalf of the practice. Team members also should be instructed to report violations. You don't want to encourage snooping and tattling, but you do want to do your best to protect the practice and let the team know that some behaviors won't be tolerated.

If your practice is using social media as a marketing tool, you must have and distribute guidelines on who can post, what they can say, and how it should be said.

Can I set standards for professional appearance and attitude among the staff?

The short answer is "Yes"; the longer answer is "It depends on how you go about it."

Begin with identifying what "standard" you want to see in your employees. Instead of attitude, however, address the actual behavior of the employee. For example, your standard could be for everyone to contribute to a productive team. If someone is not living up to that expectation, assess what they are doing that can be observed—are they leaving messes for other team members? Are they not easily located when help is needed? Do they participate in team meetings along with everyone else? You can measure these behaviors, while it is too subjective to say, "This person is not a team player."

The ability to get along with others, provide a positive client experience, treat other team members with respect, and generally present a pleasant attitude are considered soft skills. These are some of the most important traits we want in our team members, so we should specify these as characteristics to look for when hiring new team members. Include the soft skills valued by the leadership of the practice in the job description used during interviewing and hiring, and made a factor in training. Show a new hire both what to do and how to do it—with a smile, a kind word, open body language, appropriate eye contact, a tone of voice appropriate for the interaction. Make sure that performance reviews reflect those soft skills and whether they are adequate.

An image and appearance standard would typically be addressed in a personal grooming policy and dress code, which may depend in part on your clientele and region, and time of year—a more casual look may be appropriate in some communities or during the summertime, for instance.

As an employer, you want the practice team to present an acceptable image at all times, however. That is why you may want to have employees wear uniforms, which they will be more likely to do if the practice provides the outfits and covers the cost of laundering them. You also may want to include policies about tattoos and piercings, if you think such adornments might put clients off.

The Society for Human Resource Management (SHRM) provides a template that you can tailor to your practice to set and maintain a dress code. Make it part of your employee handbook and be sure to feature it prominently in your welcome to all new employees.

RESOURCES

Lee, Fred. 2009. *If Disney Ran Your Hospital: 9 1/2 Things You Would Do Differently*. Bozeman: Second River Healthcare Press.

Society for Human Resource Management (SHRM). http://www.shrm.org. Contains templates for dress codes for SHRM members.

CHAPTER 10: MEETINGS AND COMMUNICATIONS

Good and thorough communication is vital to the health and success of a veterinary practice. With different positions, different shifts, and different personalities involved, it is a constant job to keep the team updated and on the same page. You can use various methods of communication, and it makes sense to incorporate several methods to disseminate the same information and catch the attention of everyone in the practice, since different people learn in different ways.

Team meetings, perhaps the most common method of communicating with all team members, can be effective, but challenging to schedule so everyone can participate. These meetings need to be educational, informative, and fun, engaging team members so they are comfortable with contributing to discussions and ideas.

How do we get our employees to buy into and embrace our vision?

It is the responsibility of a leader to create a compelling vision and live that vision as an example to follow. Once you have established the vision for your practice, communicate it not only in writing but through stories about actions that emphasize and promote it. Describe obstacles that someone faced and actions taken to overcome them. Talk about the sense of purpose a team member got from serving a client or sick pet. Share client stories, and trials and tribulations that ended with thank-you notes of praise to show the team how the practice's vision is thriving in their everyday encounters with clients.

Involve the team in keeping the vision alive and growing by helping them see WIIFT (what's in it for them?)—how they benefit as much as the business and the pets you serve. Many employees need to see how the vision meshes with or supports their needs and their careers. If the vision helps them grow as the business grows, there is a win-win for everyone involved.

Share the vision at team meetings. Work at creating a culture that will energize the team rather than bogging them down with bureaucratic rules and regulations. Make the vision fun and let the team be free to excel (even at the risk of some failures).

If the vision was created years ago, it is probably time both to communicate it to newer team members and consider refreshing it. If the vision is due for a change, get the team to play a role in providing ideas and telling stories that are meaningful to them in terms of the purpose of their work. Above all, communicate the vision and lead by example.

Do small- to medium-size clinics need hierarchies?

Typically, a hierarchy is presented in an organizational chart for the practice. It does not "rank" employees on the basis of skills, personality, or contributions to the team. Its main purpose is to document the flow of responsibility, communication, and reporting that should occur in a practice. This communication will differ in practices that are small, medium, and large.

Some people cringe when they think about an organizational chart and are cautious about a hierarchy format. It can help to lay the chart on its side horizontally, instead of up and down vertically, so it does not give a visual impression of some people being at the "top" and others at the "bottom." The chart is more a communication conduit that demonstrates whom to go to if you have a problem or suggestion.

In a small, one-doctor practice with only a couple of support team members who are cross-trained to do many things in the practice, there may be no need for an organizational chart or hierarchy. The owner is typically in charge, and that's usually well understood.

Things start getting more complicated as the team grows, however, and the practice may experience some growing pains.

As the practice gains success, the owner may decide to bring on a part-time associate veterinarian. If the phone is also ringing at a frequent, and welcome, pace, it may be necessary to have someone focused on answering the phone and greeting clients. While one member of the team goes up front, the other support team member can focus on the medical aspect of the practice.

When there is a group of around 10 people, several team members are probably working at the same job. A few team members are up front, and a handful of medical support team members are in the back. As these two teams grow, it will be necessary to create supervisor or lead positions to ensure consistent communication and reporting. The more the business grows, the more it becomes necessary to ensure proper communication from the boss to the team, and vice-versa.

Depending on the number of employees, a small practice probably has no need for this chart, although it might be a good idea to get the team used to seeing one. When you have about 10 people or more, it is likely time to bring in a management space on the organizational chart. This also helps prevent the owner from having to deal directly with several people on a daily basis. In its essence, the chart is meant to make the future less complex.

How can I make employees understand the meaning of "mandatory" (e.g., "Staff meetings are mandatory")?

If "mandatory" is attached to something, that word must hold weight in the eyes of the team. Mandatory only works if it is enforced, equally, with every person on the team. Before you make anything mandatory, be sure you are ready to hold the team accountable for that activity or behavior—and that it is important enough for that distinction. Then, enforce the mandatory status with differing degrees of discipline.

If team meetings are to be mandatory, these factors should be considered in creating the policy you want:

- Have I arranged the best possible time for this event, to aim for success?
- What if someone comes in late?
- What if someone calls to say they can't make the meeting? Does it matter why they can't make it?
- Can absence from the team meeting be approved if it is requested ahead of time? How far ahead of time? What reasons are acceptable?
- What if someone misses the meeting without advance notice?
- Am I willing to discipline my best employees?
- How will I maintain communication so employees are aware of the issue?

It is important to aim for success. Don't just say a meeting is mandatory; explain why it is. Do not assume that calling one team meeting mandatory means employees will understand that they are all mandatory.

Schedule the event well ahead of time so employees can make arrangements to attend. Add the event to the work schedule, so they see that attending the meeting is as important as showing up for a shift. If possible, and necessary, schedule meetings at several times to accommodate everyone's schedules.

It may be necessary to have some team members on the floor during a meeting, handling clients and patients if the practice can't be closed entirely. Know who is working so they are not assumed to be absent, and try to rotate so no one misses two consecutive meetings because of a shift.

To make something mandatory for your team, you have to communicate your expectations and then hold the team accountable for being at the mandatory event. The first step is yours.

What are the best times to schedule staff meetings?

As managers and leaders, we know it is beneficial to have meetings with our teams and within our teams. These meetings are vital for disseminating information, discussing issues and successes, relaying news and concerns, and building camaraderie among team members. As important as they are, these meetings are also one of the more difficult things to schedule, depending on the type of practice, hours of operation, size of the team, and various other considerations. There is no magic answer for every team, which is probably why your practice has tried many strategies for how to hold and manage these meetings. The leadership of the practice must feel these meetings are absolutely necessary.

Some businesses schedule all-team meetings for times when the practice is closed; some even close during a lunch hour or an hour earlier than usual so everyone can attend. If you schedule a general team meeting for times above and beyond regular work hours, you may need to compensate team members for the extra time. If you don't, be prepared for resentment and, possibly, absenteeism.

Closing the practice during operating hours to hold these meetings may work beautifully for small- to medium-size day practices, but wouldn't work for an emergency facility that operates 24 hours a day, 7 days a week. The general practice may have operating hours that allow the team to meet before or after the practice is open, yet a facility with 24-hour care of patients has team members on different schedules throughout the day and night, so it may take some brainstorming and flexibility to make these meetings work.

We want team members to become engaged and contribute to the discussions during these meetings. To encourage them, you have to make them comfortable. Team meetings certainly attract more team members when there is literally food on the table. Schedule these

212

all-team meetings during lunch time, or dinner after the doors are closed, with the practice providing the food. This involves expenses on the part of the practice, but it is often a small price to pay for the ideas and team spirit that evolve from these meetings.

Smaller groups within the team can meet more frequently, especially task forces that are focused on specific issues, such as a website committee or training team.

Too often, we stumble over getting these meetings to happen, and get further away from their purpose until we stop them altogether. Fight to find the right method for your practice—your team depends on it.

97

What are some ways to make our staff meetings more engaging and productive?

When you have a meeting with the team, whether it's an all-team meeting that everyone is expected to attend or a departmental meeting specific to a certain area or position, create an atmosphere where team members want to contribute their thoughts and ideas. This means encouraging interaction. If they help bring new ideas and solutions to problems, use games to make meetings more fun and interesting (see Question 98).

If the management has information to disseminate that does not need an open discussion, do this at times and in ways other than an in-person meeting; your team members don't want to simply be talked "to" or "at." Some information can be shared through inter-office email, employee newsletters, bulletin boards, or the practice's website or intranet. Just be sure to make everyone accountable for staying up to date.

Productive interaction will only be possible if team members feel comfortable. This involves the entire culture of the practice. In a culture where contributions from team members are considered, welcomed, and acted upon, discussion will be forthcoming and people will want to participate.

Put together an agenda for each meeting and keep the session on-topic. In creating the agenda, be sure to consider the team members' needs. Before the meeting, solicit additional agenda items directly if you have a small team, or from the supervisor of each area, to make everyone feel their participation matters. Designate some time for collecting feedback and problem-solving. Set goals together and celebrate success.

Interaction can also mean that team members are responsible for providing some of the content of the meeting. As long as a topic applies to everyone, a team member can develop a short program on a specific topic of interest or recent item of concern. It brings new life to the meeting when others get to take center stage.

To set the stage at the beginning of the meeting, ask each person to contribute one good thing about the past week, whether work-related or personal. This can help break the ice and reduce tension in getting started.

Meetings cost time and time is money—so why not plan your meetings and make them a positive return on investment? Try these tips from peers and other professionals:

- *Purpose:* What is the purpose of the meeting? What are you trying to achieve? What is the goal? What end result do you hope to see?
- *Timing:* Does the meeting need to be once a day, such as a quick review for the appointments for the day; once a week to discuss activities and game plans; or once a month for guest speakers or training sessions?
- *Agenda:* Always have an agenda, to focus the meeting conversation. Distribute it ahead of time so people come prepared (have reports ready, know what should be accomplished).
- *Who:* Involve only those directly involved in or affected by the agenda.
- *Place and length:* Save time by having the room prepared and by starting and ending on time. State the purpose of the meeting, stay on course, and wrap up by restating any action steps agreed upon.
- *Summarize:* After the meeting, distribute a summary of discussions and detail any commitments. Depending on when the next meeting is scheduled, you may need to send additional memos to keep people on target.

To prolong the positive effects of these meetings, ask each team member to write down two things they felt were the most important aspects of the session, and why they feel that way. Ask for this document by the end of the day the meeting occurred.

How do I make staff meetings more educational and interesting?

No rules say that meetings have to be boring—depending on the subject matter, playing team-building games or having some fun with the discussion helps participants stay engaged, makes meetings more interesting, and even can enhance retention of information. Meetings also can be a time to have fun together, as long as the team doesn't see a light-hearted approach as wasting time. Whether the topic is new knowledge about a medical topic or a way of improving or maintaining relationships, games can provide a great environment for learning to take place as a group. Small rewards can be used for those who participate by asking or answering a question.

Try to include training or educational topics in every meeting—news about trends, treatments, medications, and techniques. Consider incorporating demonstrations and quizzes to enliven sessions.

Some organizations have rotating team roles for the meetings—different people setting training topics, reviewing previous meeting action plans, taking charge of food. Change these roles with every meeting to give everyone a chance to take the lead.

What makes you want to attend a meeting (see Question 97) for more on making meetings more productive)? What makes you feel that a meeting was worthwhile? It is probably the same for all your employees. Some strategic points to consider include:

- Send out an agenda.
- Ask for suggestions for meeting topics (and let employees research ideas and present them).
- Let employees participate—don't let management do all the talking.

- Ask for input and respect the feedback without judging.

Don't make a meeting into too much of a circus or a pep rally with not enough attention to the business at hand, but don't make the meeting all about problems, either.

How do we find time to meet regularly with team members for education, housekeeping, team-building, discussing protocols, etc., without affecting clients?

If something is important to the life and profit of the business, you will make the time. Build essential meetings into your daily, weekly, or monthly schedules. Convey a sense of urgency and importance so others do not feel they can skip out and miss a meeting, arrive unprepared, or waste time.

To find the best time for regular meetings, consider your slow times for when employee attendance can be highest. It might even be feasible to hold meetings at an outside location—it adds to the sense of importance and can reduce time constraints. It also protects participants from being interrupted by phone calls, intercom messages, or a knock on the door so everyone can focus on the purpose of the meeting.

Take advantage of social media to involve team members who cannot attend—email or tweet an update or summary, which can kick up the sense of importance. Provide links to educational material, updates, or signature pages to reinforce the commitment that is expected.

Many of us need to stop considering meetings as "necessary evils" and realize that the real "evil" is our inability to prioritize our goals and communicate effectively and efficiently. Spending some time with a mentor or coach can help the owner or manager get a handle on meetings, agendas, and priorities so those activities become more interesting and worthwhile.

100

How do I keep everyone informed of what's happening?

Keeping everyone up to date on everything going on in the practice can be a challenge, so we must consider all the methods and use what will work best for us and our team. Team meetings were designed to be a time to disseminate information; however, the time might be better spent on engaging and interacting with our team.

Consider using social media and other non-meeting mechanisms to relay mundane but important information to the team, and save meetings for more interesting exchanges. Document meeting results and ask for sign-offs so everyone is responsible for having the same information.

One method would be inter-office email, delivered with read receipts so it can be tracked when employees open messages. You also can use practice management software (PMS) options for inter-office communication. If you use this type of digital dissemination, you may have to require everyone to check for messages at least twice a day, at the beginning and end of his or her shift, and make sure they know that they are accountable for this expectation. You could also set up a log, either on the computer or in a notebook, where the employees leave notes for other team members about job-related information. Every employee also should have a physical "inbox" for notices and other paper documents.

Make team leaders or supervisors responsible for passing information along to their teams. The supervisor is responsible for knowing which of their teammates do not attend a meeting, and making the effort to individually inform those team members of what was said at the meeting. To ensure that the team members are held accountable for knowing the information, create a signature page for each individual or the entire team to sign.

You are better off communicating the same information in several ways, because different methods attract different people. Some are computer-savvy and will appreciate information in a digital fashion. Others like to have paper in hand so they can read and touch the information. To change things up, consider videotaping yourself delivering the information.

101

What is the best way to communicate or disseminate new information?

New information is exciting and inherently interesting simply because it's new. Decide who will be the best person (based on their position on the team) to handle unveiling this new information, determine who needs to hear the message, and decide on the most effective method of communication. If everyone needs the information, it can be a good idea to inform the entire team at a team meeting, but consider having smaller groups meet separately to provide a better opportunity for questions and comments.

If the practice purchases a new piece of medical equipment, such as a device to monitor patients under anesthesia, the medical team needs to hear about it, along with the technician supervisor. You might hold a larger meeting with the entire medical team to introduce them to the machine, but smaller groups of three or four may meet for hands-on training. The front-desk team needs to know as well, so they can give clients the right information about services and technology when asked. Whoever handles your website also needs to know so the news can be posted and shared with the larger audience outside the practice itself. All of these audiences need different types and amounts of information, so different ways of providing it—in writing, by email or text message, at meetings, etc.—all are worth using.

To introduce new equipment, products, services, and the like, provide written information about it on paper, through in-house memos and newsletters, and make sure it is easily referenced on the computer as well. Plan to test the appropriate employees on the new material to ensure they receive and understand the information as needed.

Be sure to include the protocol for this new item in future training manuals.

APPENDIX

CONTRIBUTORS

To those who contributed to this book, we thank you:

Stephanie Adamson, Gentle Touch Animal Hospital, Denver, CO

Charlyce Altom, North Hills Veterinary Clinic, Reno, NV

Stephan Aregger, DVM, Kleintierzentrum Räzlirain, 3254 Messen, Switzerland

Pat Kennedy Arrington, DVM, CVFP, Jefferson Animal Hospital and Regional Emergency Center, Louisville, KY

A. Babcock, Hudson Highlands Veterinary Medical Group, Hopewell Junction, NY

Judi Bailey, Loving Hands Animal Clinic & Pet Resort, Alpharetta, GA

Reba M. Barley, Linden Heights Animal Hospital, Winchester, VA

Candace R. Benyei, PhD, Schulhof Animal Hospital, LLC, Westport, CT

Pam Bergeron, RVT, CVPM, Redwood Veterinary Hospital, Salt Lake City, UT

Debbie Blanton, Charlotte, NC

Deanne Bonner, RVT, CVPM, Santa Rosa, CA

Lock Boyce, DVM, Boyce-Hollan Veterinary Services, Stuart, VA

Karen Bracken-Penley, Great Falls Animal Hospital, Great Falls VA

Lori Brownson, Scott County Animal Hospital, Eldridge, IA

Joey Bryant, DVM, Practice Owner, Plum Creek Veterinary Hospital, Kyle, TX

Andrea Bundesmann, Huntington Veterinary Hospital, Inc., Monrovia, CA 91016

Miriam Cabrera, Veterinary Skin & Allergy Specialists, PC, Englewood, CO

Shannon Cameron, RVT, Brentwood Veterinary Hospital, Brentwood, CA

Debra Carter, Hill Country Animal Hospital, San Antonio, TX

Karli Carter, Payson Family Pet Hospital, Payson, UT

Carol A. Chaney, Acacia Animal Health Center, Escondido, CA

Dena Chiddister, Dundee Animal Hospital, Dundee, IL

John D. Clark, DVM, DVM, Phoenix, AZ

Paul Cook, DVM, Merced Veterinary Clinic, Merced, CA

Kaleia Corbin, CVT, BAS, Atlantic Animal Hospital, Ormond Beach, FL

Stacy Corwin, Halifax Veterinary Center, Port Orange, FL

Lynne Dagan, Northwest Animal Hospital, Colorado Springs, CO

Ed Dartz, Arbor View Animal Hospital, Valparaiso, IN

Anna P. Davies, DVM, MS, DACVIM, Crossroads Animal Hospital, Ltd., Burnsville, MN

Shirlei Davis, Animal Care Center of Green Valley, Green Valley, AZ

Cindy Dawson, Edgewood Animal Clinic, Lakeland, FL

Katie Delmar, Animal Hospital of Tiger Point, Gulf Breeze, FL

Melissa Detrick, Honey Brook Animal Hospital, Honey Brook, PA

Charlotte D. Dietz, DVM, Brandon Animal Hospital, Roanoke, VA

Chris Duke, DVM, Bienville AMC, Ocean Springs, MS

Julie Dyer, DVM, Lakewood Animal Health Center, Lee's Summit, MO

Diane Eigner, VMD, The Cat Doctor, Philadelphia, PA

Leslie A. Epstein, CVT, BSM, Animal Emergency Clinic, Oakdale, MN

Barry Figlioli, Jr., VCA Hanson Animal Hospital, Hanson, MA

Matthew Ford, DVM, Scarborough Animal Hospital, Scarborough, ME

H. Francis, BS, CVT, CVPM, CMVH, Portland, OR

Nicole Frost, CVT, CVPM, Prescott Valley, AZ

Shayne Gardner, CVT, Coral Springs Animal Hospital, Coral Springs, FL

Martin Gilmore, DVM, MS, Mill Creek Animal Hospital, PA, Shawnee, KS

Jeff Godwin, DVM, Animal Medical Clinic, Melbourne, FL

Kathleen Gruss, DVM, Earlysville Animal Hospital, Earlysville, VA

Kim K. Haddad, DVM, San Mateo Animal Hospital, San Mateo, CA

Kendra Hanks, RVT, Dakota Hills Veterinary Clinic, Rapid City, SD

Beth Harsdorff, CVA, LCA, Atlantic Animal Hospitals, Ormond Beach, FL, and Port Orange, FL

Wendy Hauser, DVM, Coal Creek Veterinary Hospital, Centennial, CO

David Hawkins, Dogwood Pet Hospital, Gresham, OR

Dara Heidema, LVT, LVT, East Holland Veterinary Clinic, Holland, MI

Vanessa Hernandez, Animal Medical Center of Southern CA, Los Angeles, CA

Janice Herrild, Town & Country Veterinary Clinic, Marinette, WI

Mary Hester, PetMed, Antioch, TN

Angela Hoffman, Austin Vet Care, Austin, TX

Judi Jackson, Centerville Animal Hospital, PC, Snellville, GA

Paul Jensen, CVPM, SPHR, Affiliated Emergency Veterinary Services, Eden Prairie, MN

Malcolm Johnson, Washington Square Veterinary Clinic, Petaluma, CA

Susie Jones, Flower Valley Veterinary Clinic, Rockville, MD

Michele Jourdan, AVA, Clyde Park Veterinary Clinic, Wyoming, MI

Michael Joyner, DVM, East Lake Vet Clinic, Killeen, TX

Janine Kakoyannis, Northeast Veterinary Hospital, Cornelia, GA

Stith Keiser, Cheyenne, WY

Barb King, Whitney Veterinary Hospital, Peoria, IL

William S. Koch, DVM, Tuttle Animal Medical Clinic, Sarasota, FL

F. Kocher, DVM, Ocean View Veterinary Hospital, Pacific Grove, CA

Kevin Kohne, DVM, Arnold Animal Hospital, Arnold, MO

Libby Kramer, CVPM, McKenzie Animal Hospital, Springfield, OR

Emily Langsdon, Noah's Animal Hospitals, Indianapolis, IN

Susan Lassiter, CVPM, All Creatures Animal Hospital, Dunwoody, GA

Tammy Ledford, Noah's Ark Veterinary Hospital, Williamsburg, VA

Marie Leslie, DVM, Caring Hands Animal Hospital, Lake City, FL

Vickie Martin, Skyview Veterinary Hospital, Billings, MT

Joni McCracken, AHT, Dewinton Pet Hospital, Dewinton, Alberta, Canada

Jessica McDonald, CVT, VTS (ECC), Affiliated Emergency Veterinary Service, St. Cloud, MN

Christeen McKeon, North Center Animal Hospital, Chicago, IL

Christy White McLean, Animal Hospital of Waynesville, Waynesville, NC

Ken McMillan, DVM, Pell City Animal Hospital, Cropwell, AL

Marie E. McNamara, MBA, CVPM, New Hartford Animal Hospital, New Hartford, NY

APPENDIX

Marlene McPartlin, Broadway Oaks Animal Hospital, San Antonio, TX

Karen Miller, Woodland West Animal Hospital, Tulsa, OK

Jernel Miyamoto, Aloha Animal Hospital, Honolulu, HI

Laura Monahan, DVM, DVM, Atlantic Veterinary Hospital, Seattle, WA

Beth Montoya, CVPM, Pembroke Veterinary Clinic, Virginia Beach, VA

DeNelle Moore, Belle Mead Animal Hospital, Hillsborough, NJ

Karina Moser, MA, RVT, Charlotte Street Animal Hospital, Asheville, NC

Denise Mullen, Pittsfield Veterinary Hospital, Pittsfield, MA

Katherine Murphy, Companion Animal Wellness Center (City, State)

Denise M. Nadeau, Pine Tree Veterinary Hospital, Augusta, ME

Alex O'Roark, DVM, Elyria Animal Hospital, Elyria, OH

Kelley Parrish, Park Cities Animal Hospital, Dallas, TX

Chantelle Percy, Scottsdale Veterinary Hospital, Surrey, British Columbia, Canada

Stephen Pittenger, DVM, Memorial-610 Hospital for Animals, Houston, TX

Janna Poll, MBA, SPHR, VRCC, Englewood, CO

Diane Power, RAHT, South Peace Animal Hospital, Dawson Creek, British Columbia, Canada

Erica Racz, Parkview Veterinary Hospital, Monterey, CA

Carolyn Radding, VMD, VMD, Freeport Veterinary Hospital, Freeport, ME

Norm Rappaport, Animal Clinic of Morris Plains, Morris Plains, NJ

Missy Richards, Arbor Hills Veterinary Centre, Plano, TX

Donna Rider, MBA, Vestal Veterinary Hospital, Vestal, NY

Lisha Riley, LVT, Southgate Veterinary Hospital, Fargo, ND

Marcus Roeder, CVPM, Dublin Animal Hospital, Colorado Springs, CO

Gayle Rosenthal, North Dallas Veterinary Hospital, Dallas, TX

Julie Rowan, Carver Lake Veterinary Center, Woodbury, MN

Allison Royer, PHR, Liverpool Village Animal Hospital, Liverpool, NY

Stephanie Ruggerone, Animal Care Clinic, San Luis Obispo, CA

Tracey Ruzicka, RAHT, Bow Bottom Veterinary Hospital, Calgary, Alberta, Canada

Laurie Salazar, Desert Veterinary Clinic, Richland, WA

Christine Scarborough, AS, RVT, Clairmont Animal Hospital, Decatur, GA

Chris Schumacher, CPA, Cedarburg Veterinary Clinic, Cedarburg, WI

Sharon A. Seely, Riverview Animal Hospital, Riverview, New Brunswick, Canada

Dean Severidt, Pet Doctors of America, Jacksonville, FL

Julie Sheehy, AHN Animal Hospital Services, Nashua, NH

Michael J. Silkey, DVM, Armistead Avenue Veterinary Hospital, Ltd., Hampton, VA

Melanie Sloane, Animal Hospital of Colorado Springs, Colorado Springs, CO

Sharon Sprouse, DVM, MS, Penasquitos Pet Clinic, San Diego, CA

Christina Stamberger, Bay Cites Veterinary Hospital, Marina del Rey, CA

Diane Storey, CMA, Abbeydale Animal Hospital, St. Thomas, Ontario, Canada

Lolita Taylor, Healthy Pets of Rome Hilliard, Inc., Hilliard, OH

Wendy Thompson, CVT, Parkway Veterinary Hospital, Lake Oswego, OR

Melissa Thornton, CVT, Northgate Pet Clinic, Decatur, IL

Roberta Tipton, RVT, BA, MA, Vanderhoof Veterinary Hospital, Altadena, CA

Diana Upson, Marina Hills Animal Hospital, Laguna Niguel, CA

Lisa Walters, Lakeside Animal Hospital, Plantation, FL

Cheryl Waterman, CVPM, Cat Clinic of Johnson County, Lenexa, KS

M.L. Westfall, DVM, Hudson Road Animal Hospital, Woodbury MN

Tracy White, VCA Marina Animal Hospital, Venice, CA

Jodie D. Williams, Hight Veterinary Hospital, PA, Charlotte, NC

Vicki B. Williamson, Moore Lane Veterinary Hospital, Billings, MT

Melony Wood, Pioneer Animal Hospital, Leavenworth, KS

Paula J. Yankauskas, VMD, Lamoille Valley Veterinary Services, Hyde Park, VT

Note: These names were copied directly from the responses provided. Obvious errors in city or state names and hospitals have been corrected and entries have been edited for consistency. Contributors who only provided first names or preferred to remain anonymous are not listed.

ABOUT THE AUTHORS

Katherine Dobbs, RVT, CVPM, PHR, began her career in veterinary medicine by becoming a registered veterinary technician in 1992. Since then, her love of animals and the veterinary profession has led her along a path toward practice management and human resources, yet she is never far from her roots as a technician. She has moved into consulting to help technicians and all veterinary professionals discover or maintain career paths that are both personally satisfying and professionally successful. Katherine's passions outside of the profession and her nine pets include spending time with her family, and exercising her creativity with scrapbooking, jewelry making, and painting.

Louise S. Dunn has been in the trenches of veterinary medicine and practice management for more than 30 years. She spent 21 years in practice and 16 in management before combining that experience with the skills of building teams to create a consultant firm, Snowgoose Veterinary Management Consulting. She is vice president of the Veterinary Emergency Specialty Practice Association and a founding member of Vet Partners, among other professional affiliations, and enjoys supporting practices and coaching them to success. Louise's passions outside of veterinary medicine include her family, kayaking, opera, and New England Patriot's football.